More Praise for *Pythagoras' Trousers*

"Smart, bold, and provocative . . . . Sure to evoke even more inter-est than it does controversy." —Evelyn Fox Keller, professor, Massachusetts Institute of Technology, author of *Reflections on Gender and Science*

"Remarkable . . . . A fascinating journey through the intellectual his-tory that has shaped our current post-modern, scientific, and reli-gious culture."
—J. Wentzel van Huyssteen, Princeton Theological Seminary

"What I find really attractive and controversial about Wertheim's provocative book is her novel combination of, and proposed reme-dy for, what has generally been treated as two distinct problems: the marginalization of women in religious institutions and the margin-alization of women in research physics."
—Robert John Russell, founder and director, The Center for Theology and the Natural Sciences, Berkeley

"[Wertheim] writes vividly about the towering figures in a tradition of great 'mathematical men' . . . and about the religious debates that shaped their scientific genius." —Priscilla Auchincloss, *Physics World*

"Wertheim has written a brilliant popular science book. Her history of cosmology is the best I have ever read."
—Brenda Maddox, *The Observer* (London)

"It would be hard to imagine a more enjoyable book . . . You turn the pages as eagerly as you would those of any novel, carried along by an unusual lightness of touch and breadth of knowledge."
—Teresa Waugh, *The Spectator* (London)

"[Wertheim] feeds the interested non-specialist reader lucid (and slightly heterodox) pocket accounts of Galileo's trial, of the relation-ship between science and magic, of quantum physics and general rel-ativity . . . simplifying the complexities of 2500 years of history gracefully and without strain." —*New Scientist*

About the Author

Margaret Wertheim is an Australian science writer, now living in Berkeley, California. She holds a Bachelor of Science majoring in physics and a Bachelor of Arts majoring in mathematics and computing. She has written widely about science and society for many publications including *New Scientist, The Sciences, The New York Times, The Australian's Review of Books, 21C: Magazine of Culture, Technology and Science, Omni, Vogue, Elle, Glamour* and *Cosmopolitan.* She has also written and directed science documentaries for television, including the internationally award-winning series *Catalyst,* and *Faith and Reason,* a program about the interaction between science and religion today.

PYTHAGORAS' TROUSERS

$\#\ \varnothing\ \div\ \leqslant\ \leqslant\ \infty\ =\ +\ \pm\ \sim\ \surd\ >$

God, Physics, and the Gender Wars

Margaret Wertheim

W. W. Norton & Company
New York • London

Reprinted by arrangement with Times Books,
a division of Random House, Inc.
First published as a Norton paperback 1997
Foreword copyright © 1996 by Margaret Wertheim
Copyright © 1995 by Five Continents Music, Inc.

Library of Congress Cataloging-in-Publication Data

Wertheim, Margaret.
Pythagoras' trousers : God, physics, and the gender wars /
Margaret Wertheim. — 1st ed.
p.   cm.
Includes index.

1. Mathematical physics—History.   2. Women in science—History.
I. Title.
QC19.6.W47   1995
306.4'5—dc20   94-40095

ISBN 0-393-31724-2

W. W. Norton & Company, Inc.
500 Fifth Avenue, NY, NY 10110
W. W. Norton & Company Ltd.
10 Coptic Street, London WC1A 1PU

1 2 3 4 5 6 7 8 9 0

This book was written for my friends,
and I dedicate it to my best friend,
CAMERON ALLAN.

CONTENTS

ACKNOWLEDGMENTS

THE PEOPLE TO WHOM I OWE A DEBT FOR HELPING ME TO SEE physics in a new light are too numerous to name; the bibliography will provide a full listing of references. But I would like to acknowledge here a number of historians and philosophers of science whose work has been of particular inspiration to me: Edwin Burtt, Eduard Dijksterhuis, Evelyn Fox Keller, Londa Schiebinger, David Noble, David Lindberg and all at *God and Nature*, Fernand Hallyn, Frances Yates, Arthur Koestler, Mary Midgley, and Margaret Rossiter. Also the feminist historian Gerda Lerner.

It is not possible in two hundred and fifty pages to do justice to all the nuances in the history of mathematically based science since Pythagoras. However, there is a singular lack of accessible synthetic works that attempt to marry physics to its history, and to the social studies of science. This book is intended to help fill that gap, and I hope I shall be forgiven by the experts for the simplification that must necessarily creep into a book that spans 2,500 years and several disciplines in ten short chapters.

I would like to thank my agent, Beth Vesel of Sanford Greenburger Associates, whose belief in this book and perceptive understanding of what I was trying to achieve have been immensely helpful throughout. I would also like to thank my editors at Times Books: Betsy Rapoport and Richard Gerber, who have patiently chaperoned both me and the book through a much longer process than any of us imagined; also my copy editor, Susan M.S. Brown, who taught me some of the finer points of grammar. I thank the Literature Board of the Australia Council, for awarding me a Writer's Project Grant in 1991 to pursue research for this volume; and I thank Simon Jeffes of the Penguin Café

Orchestra, Book Soup, and Karen Robson for valuable services rendered.

No book gets written without being read by many colleagues, friends, and even strangers, who give of their valuable time to wade through drafts and make suggestions for improvement. Here I would like to thank: Evelyn Fox Keller, Paula Findlen, Sharon Traweek, Barbara Wertheim, Christine Wertheim, Denise Bigio, Edward Golub, Bruce Western, Rosie Mestel, Jennifer Steele, Ahrin Mishan, and Robin Powell. In particular, I would like to thank Keller and Findlen for their helpful and insightful comments; my sister Christine, for forcing me to be more philosophically subtle; and my mother, Barbara, for being the ideal and ever patient non-specialist reader. The writer can only be grateful that the 'selfish' gene is not, apparently, the only one in the human complement.

Finally I would like to thank my husband Cameron Allan, without whose care and support (literally) I do not know how I could have completed this book. His commitment to this project adds his name to the long list of men, throughout history, who have actively supported the cause of women in science.

This work was assisted by a writer's grant from the Australia Council, the Australian Federal Government's arts funding and advisory body.

FOREWORD

MOST CULTURES ARTICULATE THEIR WORLD PICTURE THROUGH mythology or religion, but since the seventeenth century the Western world picture has been articulated through science, above all by physics. For better or worse, this is the discipline through which our culture describes "reality". In traditional societies children learn their world picture as they grow up hearing mythological and religious stories that are part of the core knowledge of their people. Yet modern science has always been the domain of specialists and most of us receive little introduction to its "stories" either as children or adults. While it is true that other cultures also have their specialists and elite spheres of knowledge, in the "age of science" this trend has been elevated to unprecedented heights.

A clear understanding of your culture's world picture is surely a basic human need; for at its most essential, this is nothing less than the knowledge to know how you stand in the cosmological scheme. Indeed, I suggest, knowledge of a society's world picture is essential for psychological integrity within that society. Without such understanding an individual becomes, in a profound way, an outsider. As long as our culture continues to refract reality through the lens of science there is an *obligation* to make the science accessible to *everyone*. What is at stake here is not just individual sanity, but ultimately social cohesion. By binding people into the same cosmological framework, a shared world picture becomes one of the primary glues that holds communities together.

Most people today have no clear sense of the scientific world picture. Despite the fact that we are living through an explosion of science publishing, there is little about physics that is accessible to people with no background in the field. Although there are many excellent books on the subject, almost all assume a fair degree of familiarity with scientific ideas and ways of thinking – despite protestations to the contrary. Both the scientific community and the community of science writers, to which I belong, have a responsibility to do better. An increasing sense of that responsibility was what propelled me to write this book: so many of my friends kept asking for something on physics they could understand.

How to achieve this goal? How might it be possible to make physics accessible to the vast bulk of people who turn off "hard" science around the same time they start being interested in the opposite sex? How might one write for the truly uninitiated? I came to believe that one of the major obstacles here is that most writing about physics focuses only on the *answers*. People cannot make sense of answers if they do not first understand the *questions*. Solutions only have meaning if one has a firm grasp of the problems being addressed, and of why these problems matter.

Why, for instance, does it matter if the earth revolves around the sun or the sun around the earth? In most physics books, and in most classrooms, this is presented as a problem in celestial geometry: Is it the blue dot or the yellow one at the center? With virtually no sense of context we are told that Copernicus finally "solved" this problem by placing the yellow dot in the central position. To most students the whole exercise appears little more than an abstract mathematical game.

Yet the issue matters greatly. The question of whether the sun or the earth is at the center of the cosmic system is not just a matter of celestial geometry (though it is that as well), it is a profound question about human *culture*. The choice between the geocentric cosmology of the Middle Ages and the heliocentric cosmology of the late seventeenth century was a choice between two fundamentally different perceptions of mankind's place in the universal scheme. Were we to see ourselves at the center of an angel-filled cosmos with everything connected to God, or were we to see ourselves as the inhabitants of a large rock purposelessly revolving in a vast Euclidian void? The shift from geocentrism to heliocentrism was not simply a triumph of empirical astronomy, but a turning point in Western cultural history.

In order to comprehend the answers of modern physics (its theories and its laws), we must turn our attention first to the problems being addressed. Only if we understand what is at *stake* can we care about scientific questions. The key to understanding science, I believe, is to see how scientific theories emerge within the context of our culture at large. Science should be presented not as an isolated activity taking place away from the rest of society, but as a profoundly human and culturally contingent pursuit. With that in mind, my original aim for this book was to write an accessible, cultural history of physics for the non-scientific reader. In short, a book for my friends.

Along the way however, something unexpected happened. After two and a half years of reading widely across the history of physics, from Pythagoras to Stephen Hawking, I began to notice a pattern that nothing in my physics education had led me to expect. I saw that in every era historians of science had documented, the issue of God and religion kept raising his head. I began to realize that the outpouring of "theological" musings by physicists today – all the talk about "the mind of God" and so on – was not something new, but the latest manifestation of an age-old tradition. Physics, I came to see, had *always* been a quasi-religious activity.

This revelation shocked me, for like most people today I had been taught to believe that science and religion were enemies. Isn't that what the Galileo case demonstrated? Hadn't religious faith always been an impediment to scientific reason? Hadn't it always been so? On the contrary, physics must be seen as a "priestly" science, a discipline that throughout history has been informed as much by theological as by scientific inspiration. With the dawning of this realization, I abandoned the manuscript I had nearly completed and rewrote the book afresh from this perspective.

Moreover, and again to my surprise, the realization of a profound linkage between physics and religion also cast light on another question that interested me. Why is it that of all the sciences this is the one proving most difficult for women to break into? Why, after a generation of affirmative action in education, are women still so vastly under-represented in physics when they have been making great gains in the other sciences? In recent years feminist scholars of science have asked this question and a number of issues have been identified as contributing factors, but an essential dilemma remained. What I began to see was that while women had faced immense struggles in all the sciences, the "priestly" culture of physics had served as an *additional* and very powerful barrier. This too became part of the book, adding another element that I had not originally foreseen or intended.

It was precisely because I had *not* set out to address the religious dimension in physics, but had discovered it only after several years of intense reading, that I am convinced my thesis has validity. And much to my delight, the book has received a great deal of support from those in the history and social studies of science. Not surprisingly perhaps, my thesis has also attracted a certain degree of hostility. Some people feel affronted by the suggestion that physics, this apparently most practical of sciences, might have a theological lining. Others, while accepting that this may have been the case in the past, cannot accept that physics today has any residual overlap with religion.

It's not difficult to see why someone would think that way. In one sense physics produces the epitome of concrete, practical knowledge, knowledge that "works". Because of the phenomenal technological spinoffs of physics in our century, there is a not unreasonable tendency to assume that the desire for useful byproducts must always have been the driving force behind this science. And certainly, the desire for better machines and weapons *has* been a factor behind its evolution. But, important though that desire may be, it cannot be the full story behind the rise of modern physics.

The reason here is simple: Not until the nineteenth century did physics prove itself in any significant way as a source of practical technologies. The first major achievement in that direction was the science of thermodynamics, which led to major improvements in the design of steam engines and thereby enabled the second phase of the Industrial Revolution. Before that time, physicists had produced knowledge which itself might be useful – such as the laws of motion and gravity which allowed people to predict how the planets would move through the heavens or how cannon balls would fly through the air – but there had been little in the way of concrete technological spinoffs. Yet mathematically-based physics emerged as a mature science in the seventeenth century, two hundred years before. The coming into being of this science in the absence of technological byproducts suggests that this was not the only factor driving its emergence.

Science is not just driven by practical needs and desires, but also by psychological needs and desires. If we are to understand the evolution of physics we cannot just look at the history of its theories, we must also look at the psychology of its practitioners. We must go beyond intellectual history and also explore the emotional forces impacting on the community of physicists.

The thesis of this book is that a major psychological force behind the evolution of physics has been the *a priori* belief that the structure of the natural world is determined by a set of transcendent mathematical relations. This is a scientific variant of what is known as Platonism. As we shall see, long before such Platonism had any practical support it was justified by appeal to religious ideas. Psychologically, the emergence of a mathematically-based physics was linked to the notion that God himself was a divine mathematician. The physicist was always a kind of priest. While it would be foolish to claim that this religious impulse was the only factor behind the evolution of physics, there is no doubt it has played a powerful role. We do physics no disservice by acknowledging that. Indeed, as I hope the reader will agree at the end of the book, by delving into this fascinating psycho-social dimension in physics, we enrich our understanding of the science itself.

Just as I am not claiming that a religious impulse is the entire raison d'etre behind physics, neither am I claiming it is the full explanation for the gender inequity in the field. As in all the sciences, women face many barriers to a career in physics and the problem cannot be reduced to any one cause. What I *am* claiming is that the age-old link between physics and religion has set up powerful psychological and cultural resonances in our society that continue to serve as a barrier to women. Recognizing this barrier is no insignificant matter, for unless we understand the historical inertia of psycho-social forces we will never be able to overcome them. No amount of affirmative action will enable us to equalize women's participation in physics if we do not also address deep-seated patterns of acculturation which turn women and girls away. In speaking of the priestly culture of physics, it is not my intention to lay at its feet the whole of the problem of women's under-representation in physics, but rather to alert those who care about equity in science to an issue that urgently needs our attention.

The priestly culture of physics is not only a matter of concern for women, but for all those who care about science. In a time when so many physicists are talking about God, and when quasi-religious arguments are increasingly being brought to bear in discussions about the funding of expensive physics projects, such as particle accelerators and deep-space telescopes, there is a pressing need for some clarity about the relationship between physics and religion. The public discourse on this subject has largely been dominated by physicists themselves, who generally have little understanding of religion and often little knowledge of the history of science. In presenting a coherent account of the historical relationship between physics and religion I hope to contribute to a firmer foundation on which we may understand and critique the "theologizing" in physics today.

When one sets out to cover 2,500 years of history in 250 pages, many details must, by necessity, be glossed over. Every small section of this book has commanded whole volumes on its own. And rightfully so. I am deeply indebted to the many historians who have devoted years of their lives to the subtle detail of each era. I make no pretense to doing justice to any era and I am painfully aware of what has been left out of this book. That said, I stress that the purpose of this book is to present a broad historical overview, which is a very different goal. It is easy for experts to pick holes in any broad account, but in an age of increasing specialization, I believe there is a tremendous dearth of good synthetic overviews about science. This book makes no claim to be the definitive story of physics; it attempts to provide an accessible and non-trivial account of a subject that has too often been characterized by incomprehensible expertise and inaccessibility. Since I was a child, physics has been the intellectual love of my life; above all my aim here is to make a subject I love available to a much wider audience.

Margaret Wertheim
New York, October 1996

A NOTE ON TERMINOLOGY

In this book I use the word "physics" in a broader sense than some experts will like. I enfold into the term what are often thought of as two separate disciplines: physics and astronomy. The reason behind this elision is that the point of this book is to trace the history of western culture's attempts to describe the world in mathematical terms – a goal that, since the seventeenth century, has been the hallmark of the science known as "physics". Historically the heavens were the first part of the world to be subjected to mathematical treatment, and only with the scientific revolution was the mathetical method also extended to the terrestrial domain. Before then, the science of "physics" had nothing to do with mathematics but was based on Aristotelian thinking. While it is true that many universities today have separate departments of physics and astronomy, it is also true that since Newton cosmology has largely been in the hands of physicists. Given that the unofficial goal of physics today is to discover a universal set of equations that would unify the cosmos with the subatomic world, it is, I submit, justifiable in the context of this book to treat astronomy as a special branch of the wider activity we now call "physics".

Pythagoras' Trousers

Introduction

WHEN I WAS TEN YEARS OLD, I HAD WHAT I CAN ONLY DESCRIBE AS A mystical experience. It came during a math class. We were learning about circles, and to his eternal credit our teacher, Mr. Marshall, let us discover for ourselves the secret of this unique shape: the number known as pi. Almost everything you want to say about circles can be said in terms of pi, and it seemed to me in my childhood innocence that a great treasure of the universe had just been revealed. Everywhere I looked I saw circles, and at the heart of every one of them was this mysterious number. It was in the shape of the sun and the moon and the earth; in mushrooms, sunflowers, oranges, and pearls; in wheels, clock faces, crockery, and telephone dials. All these things were united by pi, yet it transcended them all. I was enchanted. It was as if someone had lifted a veil and shown me a glimpse of a marvelous realm beyond the one I experienced with my senses. From that day on, I knew I wanted to know more about the mathematical secrets hidden in the world around me.

In this I am not alone. For the last four hundred years, physicists have been obsessed with the search for mathematical relationships in the world around us. Theirs is a unique endeavor that has cast Western culture into the light of unprecedented knowledge and power but has also propelled us to the edge of darkness. The fruits of their science have irrevocably changed the way we live—both as individuals and as a society—for modern mathematically based physics has spawned technologies that have altered the very fabric of daily life. Electric power, radio and television, the internal combustion engine, the airplane, the telephone, the silicon chip, lasers, and optic fibers are all by-products of this science. During the last century, physics-based technologies have redefined how we work, play, entertain ourselves, and communicate with one another. At the same time, these technologies have given us unprecedented destructive power: laser-sighted guns, guided missiles, supersonic fighter planes, nuclear-powered submarines, and, of course, atomic bombs. Physicists' discoveries have changed our lives as powerfully as any political, economic, or religious forces. Along with universal suffrage and parliamentary democracy, physics is one of the primary forces that has shaped modern Western culture.

Above all, the science of physics has given us a new picture of reality. Through physics we have come to believe that we inhabit the third planet of a middle-aged star on the outer edge of a spiral galaxy that is one of millions of galaxies scattered throughout a vast, and possibly infinite, universe. We believe that this universe began some 15 billion years ago in a cataclysmic fireball we call the big bang, and that it may one day end as it implodes in a reverse big bang, called the big crunch. Similarly, modern physics has "taught" us that everything in the universe is made up of particles such as protons, electrons, and neutrons—entities whose very existence is defined in purely mathematical terms. It is these mathematico-material particles that physicists now assert are the basis of life, including human life. In the modern West, matter, space, and time have all been defined in purely mathematical terms, and we have located ourselves within this framework. Rather than see ourselves in relation to mythical heroes, gods, and religious laws, we in the West see ourselves now in relation to atoms, stars, and scientific laws.

This is not to say that everyone has a clear comprehension of physicists' world picture. Far from it. But even if, as some studies continue to show, a significant percentage of Americans still believe the earth is flat, we have *all* been deeply affected by the mathematical world picture physicists have constructed. Whatever people's private beliefs, it is *this* picture that is taught in our schools and universities, that is endorsed in encyclopedias, atlases, science magazines, television programs, and newspapers. In short, whatever private world picture individuals may hold, it is physicists' mathematical world picture that is endorsed by the public institutions of our society. Furthermore, in the late twentieth century, it is increasingly in the language of mathematics that all scientists— not only in the natural sciences, but also in the social sciences— seek to describe the world. Whatever private cosmologies we may envisage, in the official corridors of epistemological power, mathematics is king.

In addition, our hopes, desires, and expectations are being shaped by physicists' conception of reality, as more and more people dream about humanoid robots, hand-held multipurpose communications systems, nuclear fusion power stations, and conversations with aliens from faraway galaxies. Around the world, teams of physicists labor to reproduce the conditions of the sun, imagining the day they will command the energy of the stars, and NASA is conducting a multimillion-dollar search for "intelligent" signals from the cosmos. Modern physics tells us these things should one day be possible; the mathematical "laws of nature" *allow* them. At the same time, physicists have told us we must reject the notion of immaterial souls and animating spirits. Similarly, they say, it is impossible to heal by correcting the flow of energy in the body or to communicate with dead ancestors. These things, all vitally important in other cultures, are deemed *impossible* by the current picture of reality that physicists have constructed. For us, aliens, robots, and rocket power are in, but ancestors, souls,

and energy channels are officially out. Western physics hasn't just
opened up new horizons, it has also closed the door on many
others. Our conception of the universe, and of how we humans
might function within it, is now filtered through the framework of
this far-reaching science, for physics doesn't simply impinge on the
material basis of our lives, it also powerfully shapes the socially
acceptable territory of the modern imagination.

Yet four hundred years ago, this science did not exist. Until the
seventeenth century, the Western world picture was determined
not by science but by religion—not by mathematical physicists but
by Christian theologians. This Christian universe was quite unlike
the one portrayed by physics today. Instead of the quasi-infinite
depths of space we now take for granted, our forebears' universe
was rather small. The earth was at rest at the center, and around it
revolved a series of heavenly crystalline spheres that carried the
sun, moon, planets, and stars. Assigned to these spheres were ranks
of celestial beings—angels, archangels, cherubim, and the like—
and beyond this hermetically sealed, angel-filled cosmos was the
Empyrean Heaven of God. This Christian cosmos, alive with souls
and animating spirits, was dismantled by the new physicists of the
seventeenth century, who installed in its place the Newtonian cosmos, which we still effectively inhabit today.

Both modern science and Christianity are, in essence, different
attempts to locate humanity in a wider cosmic scheme. Where they
differ is in what they believe that scheme to constitute. In medieval
Christianity the cosmic scheme was primarily a spiritual setting; in
modern physics it has been purely physical. This transition from a
spiritual to a *physical* cosmology cannot be seen as simply logical;
rather, it represented an enormous shift in the psychic bedrock of
the Western subconscious. The question that confronts us is, Why
did this shift occur? Why did Western culture in the seventeenth
century develop a whole new world picture?

The Christian world picture just described lasted for roughly a
thousand years, yet in the seventeenth century its credibility collapsed. It was not that the universe itself had changed but that
people had; the old explanations that had been satisfactory for centuries no longer sufficed. Why not? Why did Western minds turn in

a new direction for explanations of the world around them? And why was it that a science based on *mathematics* came to replace religion as the source of our world picture? To pose the question another way: Why did *physicists* replace theologians at the helm of epistemological power? That mathematics should have become the basis for a world picture is far from obvious, for the ancient Egyptians and Babylonians, the Arabs, Indians, and Chinese also developed mathematics, yet none went on to use it as the basis for their world pictures. What peculiarities of Western culture led to such an extraordinary development?

In this book I propose that a significant part of the answer to these questions is to be found in the religious origins and associations of physics itself. To quote historian David Noble: "However some historians might retrospectively characterize Western science as a secular enterprise, it was always in essence a religious calling." Of no science has that been truer than physics. My aim in these pages is to trace the rise of physics in Western culture as a religiously inspired enterprise. In spite of the Galileo affair, it is simply not true that physicists and theologians were at war in the seventeenth century. From the thirteenth century through the eighteenth, champions of mathematically based science consciously aligned themselves *with* the Christian churches. Even Galileo wasn't trying to break away; his most ardent wish was to gain the endorsement of the pope. As historians have shown, the idea of a long-standing war between science and religion is a historical fiction invented in the late nineteenth century.

To understand the religious affiliations of modern physics, we must go back to well before the celebrated seventeenth century. We must begin in a time long before the birth of Galileo. First in ancient Greece, and again in medieval Europe, mathematically based science emerged from a tradition that associated numbers with divinity and viewed mathematical relations in the world around us as an expression of the "divine." For fully five hundred years, from A.D. 1200 to 1700, that tradition was nurtured within the bosom of Christianity, first by Roman Catholics and later by Protestants. Although today most physicists no longer maintain formal ties with any religion, the idea that mathematical study of the universe is a "divine" pursuit still has immense cultural currency. It is precisely this view that underlies Stephen Hawking's pronouncements about "the mind of God," and that is manifest in the flurry of books about physics with the word "god" in their titles: *The Mind of God, God and the New Physics, God and the Astronomers, Does God Play Dice?, The God Particle.* Despite the supposedly secular climate of modern science, the culture of physics continues to be permeated by a deep current of quasi-religious sensibility.

When physicists themselves write about their science, they usually present it as an objective and culturally neutral progression toward Truth. But as historians, philosophers, and sociologists of science have discovered in the last few decades, science—like all other human activities—is shaped by social and cultural forces. The evolution of physics is neither inevitable nor inexorable, but depends upon culturally contingent factors and human choice. One way of understanding this is through the device of personification. Whereas during the Renaissance "reason" was often personified as the lady Minerva, we might imagine physics in the personification of Mathematical Man. My intention, then, is to trace the history of Mathematical Man as a religious being.

Conceived in the sixth century B.C., this "character" has sought throughout history to describe the world around us in mathematical terms, and in the modern age has taken on the title physicist. We will follow his development from his "youth" in ancient Greece through his "adolescence" in the Middle Ages to his ascent into "manhood" with the scientific revolution of the seventeenth century, and on into our own century where he is currently undergoing a "midlife" crisis. Like any flesh-and-blood man, Mathematical Man has evolved within the surrounding sociopolitical matrix, and he is both a product of and a contributor to that matrix. But like any flesh-and-blood man, his evolution has also been determined by his *own* choices, for like any of us, he too has been responsible for shaping his *own* path.

But what of a Mathematical Woman? Why not personify physics as female? By speaking of a Mathematical Man I do not mean to suggest that physics is any sense an innately male activity (it certainly isn't); I only wish to represent the historical fact that until the late nineteenth century, with only a few rare exceptions, physicists *were* men. Until that time, women were almost entirely absent from the enterprise of mathematically based science. Indeed, physics remains an overwhelmingly male activity. According to the American Institute of Physics, in the United States today, women constitute only 9 percent of the total physics work force and only 3 percent of full physics professors. Yet according to the Bureau of Labor Statistics, in 1990 (the last year for which figures are available), women constituted 41 percent of biological and life scientists, 27 percent of chemists, and 36 percent of all mathematicians, statisticians, and computer scientists. In the last half-century women have made tremendous strides in the social and biological sciences, and even in chemistry and mathematics, yet they remain chronically underrepresented in physics.

Again, I propose that an important part of the explanation for this inequity is to be found in the religious origins and continuing religious currents in contemporary physics. The struggle women have faced to gain entry into science parallels the struggle they have faced to gain entry into the clergy. On the one hand, women have had to fight for the right to interpret the books of Scripture, and on the other hand, for the right to interpret what was traditionally regarded as God's "other book"—Nature. Yet just as women *are* now breaking into most denominations of the Christian clergy, except Roman Catholicism, so they are now breaking into most denominations of the "church scientific," except physics. Physics is thus the Catholic Church of science. This analogy is not merely a colorful metaphor, for physics is the science whose roots are most deeply entwined with religion. As the most orthodox denomination of the "church scientific," it too will be the last to accede to female incursion.

The association between religion and mathematically based science has its origins deep in the mists of history. It goes back to the very dawn of Western culture in sixth-century B.C. Greece. At this seminal time, when the Greeks were turning away from the mythological world picture immortalized by Homer and Hesiod, the Ionian philosopher Pythagoras of Samos pioneered a worldview in which mathematics was seen as the key to reality. In place of the mythological drama of the Olympian gods, Pythagoras painted a picture in which the universe was conceived as a great musical instrument resonating with divine mathematical harmonies. It is a vision that has inspired mystics, theologians, and physicists ever since. As Albert Einstein once declared: "The longing to behold harmony is the source of the inexhaustible patience and perseverance with which [the physicist] devote[s] himself." But to Pythagoras and his followers, mathematics was the key not simply to the physical world, but more importantly to the spiritual world—for they believed that numbers were literally gods. By contemplating numbers and their relationships, the Pythagoreans sought union with the "divine." For them, mathematics was first and foremost a religious activity.

Pythagoreanism survived as a mystical cult for a thousand years in the ancient world, but like all Greek religions, it was eventually overwhelmed by the force of Christianity. Yet it was within the context of Christianity that the Pythagorean spirit would ultimately gain its greatest expression. From the time that Europeans in the late Middle Ages reembraced the learning of the Greeks, there was a steady stream of Christian champions of mathematically based science: among them Robert Grosseteste (bishop of Lincoln), Roger Bacon (a Franciscan friar), and Cardinal Nicholas of Cusa. Long before the intellectual revolution of the seventeenth century, these clergymen forged a niche for quantitative science within the very bosom of Christianity, by reconceiving the Judeo-Christian god as a divine mathematician. Together they created a Pythagorean strand of Christianity, in which the ancient idea of numbers as gods was transposed into the notion of the biblical deity as a mathematical Creator. It was out of this Christian Pythagoreanism that the science we now know as physics would eventually emerge in the age of Galileo and Newton. Although many physicists today paint the Church as their historical enemy, in fact they owe to it an immeasurable debt.

But the same movement that reignited European interest in Greek mathematics and science was responsible for excluding women. The great revival of ancient learning that took place in the late Middle Ages was part and parcel of a program of clerical reform that made higher education available only to men training for the Church. Indeed, the medieval universities were set up to provide educational training for the clergy. Not being eligible for that career, women were excluded from academe and so were denied a role in the Christian revival of mathematical science, there being *no* other place where they could learn mathematics. When, during the seventeenth century, the new physicists ushered in a mature mathematical science, they were no more willing than their medieval forebears to open the arena of higher education to women. And so, mathematically speaking, nature remained the provenance of men alone.

Physicists today often argue that the absence of women from the field during the heyday of the seventeenth century was simply part of the prevailing sexism of the time. But that suggestion is not supported by historical evidence. As historian Constance Jordan and others have shown, throughout the Renaissance there was an increasingly strong current of feminist thought, and in the seventeenth century the idea of greater equality for women—including greater educational equality—was very much in the air. But neither the churches nor the traditional aristocracies wanted such a change in the social order, and the new scientists, for the most part, sided with these institutions *against* the more liberal ideologies that proposed a greater role for women. Modern Mathematical Man has not been just a passive observer to the tide of inequity, but all too often a willing contributor.

That inequity has continued to this day. To quote philosopher Sandra Harding: "Women have been more systematically excluded from doing serious science than from performing any other social activity except, perhaps, front-line warfare." In no science has this been more true than physics. From the seventeenth through the mid-twentieth century, women such as Émilie du Châtelet, Laura Bassi, Mary Somerville, Marie Curie, Lise Meitner, and Chien-Shiung Wu had to fight to be allowed to participate in physics at all. For most of modern history, women have not only been denied access to universities, they have also been denied entry to the professional academies where science was discussed and honored. If society in general has undoubtedly made it difficult for women to take up physics, its practitioners have far too frequently made it even harder. As we trace the rise of mathematical men we shall also chronicle the struggle faced by mathematical women. At the same time, we shall see that there have always been individual men who have risen above the fray and actively championed women: men such as Pierre Curie, David Hilbert, Gottfried Leibniz, Theon of Alexandria, and Pythagoras himself.

With the dawning of the Enlightenment in the eighteenth century, science and religion began to part company. The West entered an age of dualism in which the physical and the spiritual were increasingly decoupled. Yet, in spite of the officially secular climate of modern science, physicists have *continued* to retain a quasi-religious attitude to their work. They have continued to comport themselves as a scientific priesthood, and to present themselves to the public in that light. To quote Einstein: "A contemporary has said, not unjustly, that in this materialistic age of ours the serious scientific workers are the only profoundly religious people." Post-Enlightenment physicists' theological claims are not grounded in the traditional notion of God as the spiritual Redeemer of humanity, but in the conception of God as Creator of the material world. Physicists such as Einstein and Hawking have presented their work as a quest to illuminate the mathematical plan of Creation—a plan they imply is of divine origin.

I suggest, however, that this priestly conception of the physicist continues to serve as a powerful cultural obstacle to women. It continues to help fuel the long-standing belief that mathematical science is properly a men's activity. In an age when women *have* made great advances in so many other sciences their scarcity in physics demands an explanation, because we cannot hope to redress this inequity unless we understand the reasons behind it. By tracing the religious threads in the history of physics, and by examining those threads today, one of my aims is to shed some light on the problem of inequity in the field. Given the immense power tied up with this science, the lack of women physicists is not a matter we should lightly ignore. Most people now understand the importance of women participating in politics; far fewer understand how vital it is that they also participate in science, particularly in physics.

The importance is twofold. First, as long as women are marginalized in physics, they will not be able to play a significant role in developing the technologies that spin out of this science, or in deciding how those technologies are put to use. The silicon chip industry, the telecommunications industry, the electric power industry, the transport industry, the mining industry, and the aerospace industry all rely on the technological fruits of physics. By not being more involved in developing and implementing these technologies, women concede what is in the end a vast realm of social power and responsibility. Furthermore, without greater participation in physics, women cannot be involved in deciding which technological goals physicists will even try to realize.

Second, as long as women are not more involved in physics, they cannot play a significant role in determining the directions and goals of the science itself. This is a particularly crucial issue because in the last few decades the physics community has become almost fanatically obsessed with a goal that I suggest offers very few benefits for our society. That is the dream of finding a unified theory of the particles and forces of nature—a set of mathematical equations that would encompass not only matter and force, but space and time as well. In such a synthesis, everything that is would supposedly be revealed as a complex vibration in a universal force field. Protons, pulsars, petunias, and people would all be enfolded into a mathematical "symmetry," wherein the entire universe would be described as math made manifest. This is what physicists envisage when they talk about a "theory of everything," often simply referred to as a TOE.

Such a theory is really a quasi-religious rather than a scientific goal. It is this that Stephen Hawking has associated with the "mind of God," and that has prompted much of the recent theological outpouring by physicists. Even the most ardent TOE proponents acknowledge that this theoretical synthesis is unlikely to have *any* application to daily human life—not even for military purposes. Physicists seek the knowledge, not because they believe it has the potential to improve the concrete human condition, but simply because they yearn to see what they believe is the mathematical plan of *Creation*. The problem is, such a theory cannot be obtained by thought alone. In order to pursue their quest, TOE physicists during the last two decades have had to build increasingly expensive particle accelerators. It is the desire for a theory of everything that prompted American physicists to propose the ten-billion-dollar-plus Superconducting Supercollider. The sheer expense of pursuing this goal has thus transformed it into an issue for society at large—because it is our taxes that would have to pay for these machines. In TOE physicists' attempts to convince us that this goal is worth the vast expenditure, they have increasingly been pressing an association between it and God. In his 1993 book, *The God Particle*, Nobel laureate Leon Lederman (a major advocate of the Supercollider) likened particle accelerators to cathedrals and hinted that the deity lurks at the end of a proton beam.

I suggest that contemporary physicists' obsession with a theory of everything is socially irresponsible. In expecting society to provide billions of dollars to support this quest, TOE physicists have become like a decadent priesthood, demanding that the populace build them ever more elaborate cathedrals, with spires reaching ever higher into their idea of heaven. Since a theory of everything would be not only utterly irrelevant to daily human life and concerns, but also incomprehensible to the vast majority of people, TOE physicists can be likened to the late medieval Scholastics. This is the twentieth-century equivalent of asking how many angels could dance on the head of a pin.

I believe we need a new *culture* of physics, one that does not place so much value on quasi-religious, highly abstracted goals; a culture that is less obsessed with particles and forces, and more concerned with human beings and our needs. One of the roles I believe women might play in physics is to encourage a shift away from the present obsessions. I am not suggesting here that women are innately uninterested in theories of particles and forces, or that male physicists have inherently different interests from females, but rather that modern physics has evolved in such a way that it now tends to attract only people, of both sexes, with certain kinds of interests and proclivities. What I am advocating is a culture of physics that would encourage both men *and* women to pursue different kinds of goals and ideals.

One of the reasons more women do *not* go into physics is precisely that they find the present culture of this science, and its almost antihuman focus, deeply alienating. As do many men. After six years of studying physics and math at university, I realized that much as I loved the science itself, I could not continue to operate within such an intellectual environment. It was not that I had lost faith in the value of physics, only that I could not function in the atmosphere in which it was being practiced. Since then, I have dreamed of an environment in which one could pursue the quest for mathematical relationships in the world around us, but within a more human-centered ethos.

Let me stress, then, the problem is not *that* physicists use mathematics to describe the world, but rather *how* they have used it, and to what *ends*. There is nothing in a mathematical approach to nature that demands a focus on particles and forces, or on arcane abstraction. Because science is always a culturally directed pursuit, there is no reason that we cannot have a mathematically based science focused on different goals and dreams. Such a science would not be practiced just by women, but also by men. In fact, there are some male physicists who have already voiced strong opposition to the TOE quest. Again, the issue is not that physics is done by men, but rather the *kind* of men who have tended to dominate it. Mathematical Man's problem is neither his math nor his maleness per se, but rather the pseudoreligious ideals and self-image with which he so easily becomes obsessed. He does not need a sex change, just a major personality realignment.

I propose that one of the reasons Mathematical Man has evolved this "decadent priest" persona is that he has spent most of his history without female company. I do not mean to imply that a greater presence by women would suddenly turn physics into an ideal science, but only that women would provide—as do women in all societies—a balancing influence on the physics community, and on the practice of the science. Historian of science Elizabeth Fee has remarked that trying to imagine a genuinely female-inclusive science is "like asking a medieval peasant to imagine the theory of genetics or the production of a space capsule." What Fee and others envisage is a time when women will be equally involved not just in *doing* science, but in determining *what science is,* how it is *practiced,* how it is *put to use* in our lives, and above all, what are its ideals and goals. In no other science is such a reenvisioning so difficult, or of such consequence, as it is in physics. In the final chapter, we will ask what women might bring to physics, and how their involvement alongside men might lead to a different scientific culture. In other words, we shall consider what Mathematical Man and Woman might be able to achieve *together*.

1

$\Omega \quad \mu \quad \delta \quad \Omega \quad \omega \quad \pi \quad \Sigma \quad \Delta \quad \gamma \quad \beta \quad \alpha \quad \mho$

All Is Number

ACROSS EURASIA THE SIXTH CENTURY B.C. WAS A TURNING POINT FOR humankind. This was the century of Confucius and Lao-Tsu in China, of Buddha in India, of Zoroaster in Persia, of the Ionian philosophers, and of Pythagoras in Greece. Out of this miraculous century came the great Chinese and Indian "ways of liberation," Taoism and Buddhism, while Confucius formulated the principles of conduct that would shape the psychic bedrock of his nation. In this seminal age the tectonic plate of the Asian psyche shifted and gave rise to a great spiritual flowering throughout the East. In the West a revolution of a different sort was taking place, for this was when the Greeks began to turn away from mythological explanations of the world around them and to seek instead physical causes.

Until then nature had been seen as a drama played out by the gods: The sun rose, the rain fell, the seasons passed, and the harvest was good according to the routines and whims of supernatural beings. Apollo, the sun god, rode his fiery chariot across the sky each day; Poseidon, the god of the sea, conjured up storms to impede sailors; and farmers made offerings to Demeter, goddess of agriculture, for a bountiful crop. The world was alive with the powers and personalities of a plethora of gods and goddesses, fates and furies, daemons, nymphs, titans, and other nonhuman beings. In this mythological framework, the gods could be appeased but not predicted. Then, around 600 B.C., the idea was born that natural phenomena were effected not by gods, but by processes inherent in nature itself—processes that could be comprehended, and predicted. A mechanistic mode of thinking began to cut through the rich layers of myth as people turned to nature itself to ask: What are you?

The epicenter of this intellectual revolution was on the shores of the Aegean Sea in the cities of Asia Minor—what is now Turkey. Here the Ionian philosophers addressed themselves to discovering the mechanics of nature. In contrast to the Greek myths, which had explained the world in terms of *psychological* forces incarnated in the various gods, the Ionians sought to explain the world in terms of *physical* forces and processes. Turning away from the anthropic drama of the Olympian deities, they sought naturalistic explanations, for the Ionians believed the world was a rational system that people could discern for themselves through the power of their minds. Thales envisaged the earth as a huge disk floating on a vast ocean, while Anaximander saw it as a great cylinder floating in midair. Likewise, he saw the sun as a huge hollow wheel filled with fire that revolved around the earth. Although the Ionian visions were often dreamlike and surreal, they were the first attempts to explain natural phenomena without invoking supernatural powers. The ideas of individual Ionians have largely been forgotten, but it was from their collective vision of naturalism that what we now recognize as Western science first emerged.

Of all the Ionians, one in particular has inspired thinkers ever since. In contrast to his fellow philosophers, who concentrated on explaining the world in terms of material elements such as earth, air, fire, and water, Pythagoras of Samos saw the essence of reality in the immaterial magic of numbers. He believed the universe could be explained by the properties of numbers and the relationships between them, a philosophy encapsulated in his famous dictum "All is number." At the same time Pythagoras was a deeply religious man. Rather than abandon the traditional gods, he incorporated them into his mathematical world picture. A mystic, mathematician, and philosopher, Pythagoras fused the rationalism of the West with a mysticism he learned from the East and created a unique philosophy cum science cum religion. From the seed of his extraordinary vision would be born the science of modern physics.

Even in his own time, Pythagoras was a legend. Rumored to be the son of the god Apollo by a virgin birth to his mother, Pythais, he was said to have worked miracles, conversed with daemons, and heard the "music" of the stars. He was regarded by his followers as semidivine, and there was a saying that "among rational creatures there are gods and men and beings like Pythagoras." It is difficult to sort out fact from fiction about his life, for he lived in that brilliant but hazy zone where myth and history collide. None of his writing has survived, but ancient sources abound with references to him. Even in the works of that most logical of ancients, Aristotle, we find accounts of Pythagoras that mix tales of miracles with discussions of his mathematics and cosmology. Pythagoras' philosophy fully reflected his transitional age, for while it contained the seeds of mathematical science, it also maintained a role for the pantheon of gods. Both in thought and in life, this Samian sage was a bridge between two worlds.

In many respects the mythico-religious dimension of Pythagoras' life bears an uncanny resemblance to the life of Christ depicted in the New Testament. Both men are said to have been the offspring of a god and a virgin woman. In both cases their fathers received messages that a special child was to be born to their wives —Joseph was told by an angel in a dream; Pythagoras' father, Mnesarchus, received the glad tidings from the Delphic oracle. Both spent a period of contemplation in isolation on a holy mountain, and both were said to have ascended bodily into the heavens upon their deaths. Furthermore, both spread their teachings in the form of parables, called *akousmata* by the Pythagoreans, and a number of parables from the New Testament are known to be versions of earlier Pythagorean *akousmata*. One historian has suggested that early Christians may have taken elements of the Pythagorean myth and attributed them to their own prophet, for in the ancient world Pythagoras was known first and foremost as a religious figure. During the closing centuries of the Roman Empire, when Christianity was just another cult vying for religious supremacy, a great revival of Pythagoreanism occurred, and latter-day followers of the "Master" promoted him as a Hellenistic alternative to the "king of the Jews." As had Christ, the Samian sage had promised mystical union with the divine, and to his Roman followers his teachings offered a rational spiritual alternative to the rising tide of Christianity.

Pythagoras is known to have been born around 560 B.C. on Samos, a prosperous island in the Aegean Sea not far from the coast of Asia Minor. An important gateway to the commercial cities of the mainland, Samos was also religiously significant, being the site of a monumental temple to Hera, queen of the Olympian gods. On this island, Pythagoras was always something of an outsider, for while his mother is said to have been a native of Samos, his father was a foreigner, probably a Phoenician, who had been made an honorary citizen for giving grain to the Samians during a time of drought. As an ethnic half-caste, Pythagoras was not considered a true Greek, and furthermore, his mystical bent singled him out early as a misfit among the Samians. Later in life he turned away from Ionian culture and identified himself strongly with the East, an allegiance he symbolized by rejecting the long robes favored by the Greeks and adopting instead the Persian fashion of trousers.

As a wealthy merchant, Mnesarchus could afford to educate his son, and in this age of awakening, the young Samian was taught by some of the greatest of the new Ionian thinkers. His instructors included Anaximander, Pherecydes, and Thales, one of the legendary Seven Sages and the first true philosopher. But although Pythagoras was trained by the best philosophers of the time, he hankered for something more, and after absorbing the best of the West, he set out for the East—initially Egypt, and later Babylon. (Thales had recommended that if he wished to be the wisest man alive, Pythagoras should go to the land of the pharaohs, where geometry had been discovered.) There is controversy among historians about whether Pythagoras really made a trip to Egypt and Babylon or whether it was an invention of later disciples. But either way, historian David Lindberg has pointed out that the story encapsulates an essential historical truth: The Greeks inherited mathematics from the Egyptians and Babylonians, and Pythagoras is regarded as the person who introduced this treasure to the West. Because he was undoubtedly the first great Greek mathematician, we shall assume, along with the ancients, that the journey did in fact take place.

According to Iamblichus, his third-century Roman biographer, Pythagoras traveled to Egypt by way of the Levant, the lands bordering the eastern shores of the Mediterranean. His intention was to learn the sacred rites and secrets of the region's religious sects. Some people collect stamps, others coins; Pythagoras collected religions, and he made it his business to be initiated into as many as he could. In this some of his ancient detractors accused him of cynical motives, and even his supporters acknowledged the accusation was partly true. As a young man Pythagoras certainly aspired to a career as a public speaker, and he clearly understood the public relations value of exotic mystical experiences. Nonetheless, he was also a genuinely religious man.

When Pythagoras arrived in the land of the pharaohs, events didn't go entirely as he had hoped, for according to Porphyry, another Roman biographer, he was rejected by the priests at the temples of both Heliopolis and Memphis. Eventually, however, he was accepted at Diospolis, where he studied for many years. The ancients disagree about how long Pythagoras spent among the Egyptians, but it seems to have been at least a decade. Porphyry tells us the priests imposed harsh tests on their foreign aspirant, but what he learned from them will forever remain a mystery because Pythagoras always honored their fanatical secrecy, which he would later make a cornerstone of his own religious community.

Pythagoras' Egyptian sojourn came to an abrupt end in 525 B.C., when the Persians invaded Egypt and he was taken as a captive to Babylon. In that fabled city of the hanging gardens and the great ziggurat, he availed himself of the wisdom of the Babylonians.

According to Porphyry, he studied under the sage Zaratas, from whom he learned astrology and the use of drugs for purifying the mind and body. He was also initiated into the mysteries of Zoroastrianism, with its opposing cosmic forces of good and evil. This dualism would profoundly influence his own thinking and would eventually be incorporated into his mathematico-mystical philosophy. The Babylonians were not only astrologers but also great astronomers and mathematicians. Lindberg notes their mathematics was of "an order of magnitude superior to that of the Egyptians." From them Pythagoras may well have learned the theorem for which he is still famous today: that for a right-angled triangle, the square of the hypotenuse is equal to the sum of the squares of the other two sides. Although we are taught in school that this is the Pythagorean Theorem, historians of mathematics believe it was almost certainly known to the Babylonians before him.

If Pythagoras had been an oddity on Samos before he left for the East, how much more a misfit he must have been upon his return after two decades spent with foreign priests and sages. Now he not only wore trousers, but also never cut his hair or beard—a habit that would later become a hallmark of Pythagorean followers. On Samas, he set himself up to teach philosophy and mathematics, offering lectures in the open air. Yet it soon became clear that his mystical leanings had little appeal to the Samians, so once again he left his homeland, this time forever. Pythagoras' aim now was to found his own community, where committed followers would dedicate themselves to a life of religious contemplation and study of the "divine." As the site of this utopian community he chose the town of Croton in southern Italy—a place at the very extremity of the Greek world.

Since none of the Pythagorean community's writings or records have been preserved, because of the group's fanatical secrecy, the details of its operations remain shrouded in mystery, but we do know that the lives of its participants combined elements of Greek religious practice with Egyptian-inspired rituals. In addition, the community also operated as a philosophical and mathematical school. Members were of two kinds: the *akousmatics* and the *mathematikoi*. The former lived outside the community and visited only for teaching and spiritual guidance. They did not study mathematics or philosophy but were taught through *akousmata*, which espoused a simple, nonviolent way of life. For them Pythagoreanism was essentially an ethical system with mystical undertones and Pythagoras was a purely spiritual leader.

The *mathematikoi*, however, lived inside the community and dedicated themselves to a Pythagorean life. That life was communistic in the sense that initiates had to give up their property to the community and renounce all personal possessions. Pythagoras believed this was necessary in order for the soul to be free from extraneous worries. Inspired by his life among the Egyptian priests, he was also greatly concerned with purification, and the *mathematikoi* were not allowed to eat meat or fish, or to wear wool or leather. It was said by ancient commentators that initiates had to undergo a probationary period of up to five years, during which they were to be silent to demonstrate their self-control. While it is unlikely that silence was complete, it is clear that only those who were truly dedicated made it into the inner circle, handpicked by Pythagoras, to hear the Master's most secret teachings and study mathematics. Following the model of the Egyptians, all knowledge was kept secret within the community, and one member was expelled when he revealed the mathematical properties of the dodecahedron, one of the five "perfect" solids. Pythagoras believed that, as divine knowledge, mathematics should be revealed *only* to those who had been properly purified in both mind and body, and the *mathematikoi* approached its study in the spirit of a priesthood.

The Pythagorean community at Croton is often referred to as the brotherhood, yet this is a misnomer because it also included women. Pythagoras himself was married with several children, and his wife, Theano, was an active member and teacher in the community. But the controversial question is not so much whether women could be Pythagoreans but whether they were allowed to become *mathematikoi*, philosopher-mathematicians, or only *akousmatics*. Because no record of the community survives, it is difficult to resolve this issue, but in the writings of a number of ancient commentators there is evidence that there were women *mathematikoi*. Theano, for example, is said to have written treatises on mathematics and cosmology. The idea that women could be members of Pythagoras' inner circle is also lent credence by the fact that Pythagorean communities in the fifth century B.C. also included women: Phintys, Melissa, and Tymicha are three whose names have come down to us. Finally there is the example of Plato, who was deeply influenced by Pythagoreanism and was the only one of the great Athenian philosophers who advocated the education of women. Unlike Aristotle, Plato allowed women into his famous Academy, where mathematics was taught. Thus it seems reasonable to conclude that among the original Pythagoreans women *did* participate in mathematical study. Given the nature of Greek society at the time, it is highy unlikely there were as many women as men among the *mathematikoi*, but given how misogynist the Greeks were soon to become, the community at Croton must be seen as one of the more gender-equitable havens of the Greek world. In the beginning, then, mathematical men acknowledged and accepted the presence of mathematical women.

The last years of Pythagoras' life are clouded in shadow. Between 510 and 500 B.C., a Croton nobleman, Kylon, led an uprising against the Pythagorean community, which resulted in its demise. This event has been variously described as religious persecution and as a democratic revolt against an aristocratic sect. Ancient champions of Pythagoras depict Kylon as a tyrannical man motivated by revenge after having been rejected by the Pythagoreans, yet some historians believe the backlash against the community was a response to its elitist and secretive nature. During the uprising Pythagoras fled and supposedly spent the rest of his long life wandering in Italy, spreading his teachings. It is said that on his death he ascended directly into the heavens from a temple of the muses.

Because of the secrecy surrounding the original Pythagorean community, it is impossible to say which ideas came from the Master himself and which were developed by his disciples; traditionally all ideas were attributed to Pythagoras. At the heart of Pythagorean thought were the whole numbers: 1, 2, 3, 4, 5, and so on. (Greek mathematicians had not incorporated the concept of zero.) Pythagoras believed that numbers were divine, and he equated them with the gods. The numbers 1 through 10, those of the *decade*, were said to be especially sacred. But in equating the gods with numbers, Pythagoras radically reconceived the traditional Greek pantheon: No longer anthropomorphic beings playing out a grand emotional drama, the deities had become abstract mathematical entities. The Pythagorean world picture was not the cosmic theater of Homer and Hesiod but a metaphysical dance of numbers.

Yet this number universe was richer than most modern minds might imagine because the Pythagoreans believed that numbers had ethical and moral characteristics. Thus, as in earlier Greek mythology, their cosmology retained a *psychological* dimension. Today we see 4 as simply a quantity that allows us to say that there are four seasons in a year or four sides to a square. But to them, 4 was a great deal more than this—for instance, it was the number of justice. Four is 2 times 2, which to them it signified a balanced scale. Similarly, 6 was the number of marriage, because 6 is 2 times 3 and they regarded 2 as a female number and 3 as a male number, so 6 was the first male-female product. The idea of numbers having nonquantitative characteristics was very possibly another facet of Pythagoras' philosophy that he acquired in Egypt, for this was also a feature of Egyptian numerology.

For the Pythagoreans, numbers' nonquantitative properties meant that they could serve as ethical archetypes, and so the study of mathematics could provide insight into human behavior. Above all, because odd numbers were considered male and even ones female, the specific properties of odds and evens had moral implications for the sexes. In particular, the Pythagorean perception that odd numbers were good and even ones evil cast women definitively on the side of evil. Thus we see the emergence of Pythagorean dualism: On one side were the qualities of goodness, oddness and maleness; on the other, the qualities of evil, evenness and femaleness. As a rule, those qualities regarded as higher or better were cast on the side of the male, while those regarded as lower or worse were on the side of the female. In general, the Pythagorean mathematician's task was to discover the characteristics of individual numbers (whether odd or even), and the relationships between them—both numerical and ethical. The mathematician was thus inevitably a student of morality. The modern separation between mathematics and ethics that we so take for granted would have horrified Pythagoras, who was one of the first to understand that math could be applied to the development of destructive technologies, and hence entails a moral responsibility.

As well as serving as ethical archetypes, the Pythagoreans believed that numbers also served as material archetypes. Indeed, they saw numbers as the models for all physical form. The notion of numbers as the source of form originated in the Pythagorean discovery that each one could be associated with specific shapes. For instance, 6, 10, and 15 were called triangular numbers because six, ten, or fifteen dots could be arranged into equilateral triangles.

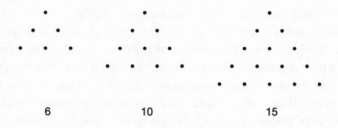

Similarly, 4, 9, and 16 were called square numbers.

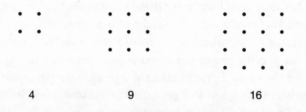

Twelve was considered a rectangular number, because it could be made up of three rows of four dots and also two rows of six dots. Some numbers, such as 6 (which is both triangular and rectangular), could take on more than one shape. Dots could be arranged into almost any shape one chose: pentagons, hexagons, octagons, and so on. Pythagoras reasoned that if numbers had forms, then perhaps, reciprocally, all forms could be associated with numbers.

Indeed, might not number be the essence of form itself? A simple example of associating a number of dots with a complex form could be seen in the constellations of the stars. Here a pattern of half a dozen or so points of light became a ram, a bull, a crab, or a man. Historians have suggested that Pythagoras may have conceived of numbers as the essence of form because of his experiences among the Babylonian astronomers.

Using dot patterns the Pythagoreans discovered surprisingly complex mathematical theorems. Furthermore, although mathematicians in later ages would use much more sophisticated techniques, the idea that form is essentially mathematical has proved immensely fruitful. Mathematicians no longer consider arrangements of dots; instead they try to describe form with equations. For example, a circle can be described by the simple equation $x^2 + y^2 = r^2$ (where x and y are distances along a horizontal and vertical axis, and r is the radius of the circle). A variation of this equation represents an ellipse, which is, in effect, an elongated circle. It turns out that the orbits of the planets around the sun take the form of ellipses, as does the orbit of the moon around the earth. Thus a single simple equation (a mathematical form) embodies the dance of the solar system.

The Pythagoreans noticed that numbers are manifest not only in *spatial* patterns but also in temporal ones. Each year consists of 4 seasons, 13 lunar months, and 365 days. The sun rises and falls in a regular cycle every 24 hours, the moon waxes and wanes every 29 days, and each of the planets has its own unique rhythm. Thus the cosmos can be seen as a grand series of numerical cycles. The temporal numerical patterns apparent in the heavens and the spatial patterns made by numbers themselves convinced Pythagoras that all was indeed number, and that number was truly the essence of reality.

Most importantly for Pythagoras, the human psyche was part of the great numerical pattern of the universe, for he believed we are all reincarnated every 216 years. Known as the psychogonic cube, the number 216 had special significance, being 6 cubed, or $6 \times 6 \times 6$. To the Pythagoreans, this symbolized cyclic return—the notion that all things repeat themselves. As had the Hindus, the

Pythagoreans believed they were fated to an endless cycle of life after life after life, and so, like numbers themselves, psyches were thought to be immortal. Indeed the immortality of the psyche and its endless cycle of reincarnation were the primary religious doctrines of Pythagoreanism—one aspect where the influence of the East was again much in evidence.

Because the psyche was continually reincarnated, according to Pythagoras, it spent time alternately on earth (trapped in a body) and not on earth (disembodied). Thus arose the question: When the psyche is not incarnated, where does it spend its time and what is it doing? Pythagoras' answer was that it dwelled in the heavenly realm of the immaterial number-gods, where it basked in bliss amid the strains of the mathematical music of the cosmos, the so-called harmony of the spheres. We shall explore this cosmic harmony in more detail shortly; suffice it to say for the moment that the Pythagoreans considered hearing the harmony of the spheres to be the epiphany of human experience. Although every person could look forward to this divine pleasure after death, Pythagoreans aimed to free their psyches to experience that pleasure while alive. Pythagoreanism then was primarily an ascent religion, in which the aim was to use mathematics as a tool to free the psyche from the body so it could rise into the "heavenly" numerical realm. Mathematics was thus first and foremost a religious activity.

Yet in parallel with this rather beautiful mysticism was another agenda, for in freeing the psyche Pythagoreans were also attempting to escape from *nature*. The reason Pythagoras equated numbers with the gods was that to him they seemed timeless, immutable, and incorruptible. In this numbers stood in stark contrast to the all-too-obvious fact that everything in the material, or natural, world is subject to corruption, decay, and death. Four flowers might wither, four melons might rot, four men might die, four streams might dry up, four mountains might crumble, but the number 4 itself seemed eternal and indestructible—like the gods. Pythagoras equated divinity with timeless stasis and immutability, qualities which cannot be found anywhere in nature. Indeed the whole point of the number-gods was that they were beyond nature, with its inherent transience and mortality. Here again Pythagoreanism marked a new phase in Greek thinking, because although the traditional gods had been considered immortal and indestructible, far from being beyond nature they were embedded in it. After all, they were directly responsible for the rising of the sun and moon, for the bounty of harvests, for storms and thunder, for oceans and wind.

But transcending nature also implied transcending the "feminine," because in Pythagoreanism, as in most Greek thought, matter—the very substance of nature—was considered inherently female. In Pythagorean cosmology, the number 2 was not only the supreme female principle but also the number associated with matter. Meanwhile the supreme male principle, the number 1, was equated directly with the supreme immaterial deity, Apollo. Here then was another central feature of Pythagorean dualism: Maleness was associated with the heavenly and immaterial whereas femaleness was associated with the earthly and material. This male-female, heaven-earth dichotomy was not of course unique to the Pythagoreans. It had been a central feature of Greek mythical cosmology since before Hesiod compiled his *Theogony* in the eighth century B.C. Indeed, historian Gerda Lerner has shown that the polarity of a "Sky Father" and an "Earth Mother" was endemic in Western mythology from Mesopotamia to Greece. In seeking to free the immaterial psyche from the material body, the Pythagoreans were seeking to escape from the realm of the Earth Mother (in their mathematical mythology, this being represented by the number 2), and to ascend into the realm of the Sky Father (represented by the number 1).

Pythagoras did not invent this male-female, heaven-earth dichotomy, he simply transformed it into a *mathematical* context. The really interesting development here was that because in Pythagoreanism all numbers, in and of themselves, were regarded as belonging to the psychic realm (that is, the male realm), mathematics as a whole came to be regarded as an essentially male activity —not in terms of who was entitled to practice it, but in terms of its inherent nature. The study of mathematics was seen by the Pythagoreans to be the study of the transcendent realm of the number-gods, and it was the supposedly male element of a human being (the psyche) which was engaged during this activity; while the supposedly female element (the matter of the body) was to be left behind when one engaged in mathematics.

This association of mathematics with maleness has had a profound impact on Western culture ever since. Two and a half thousand years later, there continues to be a widespread belief that math is an inherently male activity and that men's minds are innately more suited to mathematical thinking than women's. In the sixth century B.C., one consequence of this male-math linkage was that even though Pythagorean women were able to engage in the study of mathematics, (because they too had psyches), for them the practice entailed a profound tension. To study math they supposedly had to leave behind their femaleness and strive to become purely male—they had to attempt to become, in effect, mathematical *men*. The implication of Pythagorean dualism was that a mathematical *woman* was a contradiction in terms. Although Pythagorean mysticism has long since disappeared, in this respect it has cast a long shadow across Western culture, for even today Mathematical Woman is *still* struggling to defend herself as a legitimate figure.

The central axis of Pythagorean mysticism—mathematics, maleness, and psychic transcendence—has also had a profound impact on the culture of Western physics. Because mathematics was seen as the study of the transcendent realm of the gods, so too the quest for mathematical relationships in the physical world also came to be seen as a transcendent activity—a quest for that part of nature that was eternal, immutable, and incorruptible. A quest for some part of nature that was, in effect, *beyond* nature. When, in the Middle Ages, a Pythagorean spirit reemerged within the context of Christianity, this attitude was readily transformed into an association between nature's mathematical relations and the Christian God. Like the ancient Pythagoreans, medieval and early modern physicists believed they were discovering a transcendent, deity-driven blueprint for creation.

Although the Pythagoreans were primarily a religious cult, they were also the forerunners to modern physicists, for just as they were interested in the forms embedded in numbers, so too they were interested in finding mathematical forms embedded in the physical world. That Pythagoreanism was a true protomathematical science is apparent in Pythagoras' famous discovery of numerical ratios underlying musical harmonies. In an instrument such as a lyre, a string twice as long as another produces the same note an octave lower—the ratio of the lengths of the strings is 2 to 1. Two strings in the ratio of 3 to 2 produce the musical fifth, while two in the ratio of 4 to 3 produce the fourth. The Pythagoreans also experimented with varying the thickness and tension of strings and looked for numerical patterns in the sounds produced. Their discovery of mathematical ratios underlying the phenomena of sound tangibly demonstrated that mathematics was not simply an abstract game but that it inhered in the physical world.

The ratios behind the music were said to represent its *armonia,* the mathematical music behind the audible sound. Audible harmony thus became the sensory manifestation of this mathematical harmony—the visceral experience of the transcendent realm of number. According to Pythagoras the soul responded to this *armonia* when a person listened to music. Thus, he said, people find musical harmonies such as the octave, the fourth, and the fifth pleasing because the soul has a natural affinity for simple ratios such as 2 to 1, 3 to 2, and 4 to 3. The ear is pleased because the soul is.

The idea of *armonia* was to prove enormously influential, eventually becoming the foundation of Western cosmology. Pythagoras saw the universe as a vast musical instrument suffused with mathematical harmonies. Of special interest were those inherent in the heavens—in the arrangement and motion of the sun, moon, planets, and stars. He believed that just as there were ratios behind the harmonies of music, there must also be ratios behind the dance of the heavenly bodies—such ratios constituted the so-called harmony of the spheres, the divine mathematical music the soul would hear once it escaped the material prison of the body. From this Pythagorean quest for cosmic harmony evolved not only the ancient Greek picture of the cosmos but, as we shall see, ultimately also the modern one.

The harmony of the spheres was so called because Pythagoras

believed the cosmos was spherical in shape. In his cosmology, each of the celestial bodies traveled in great circles that delineated the diameters of a series of concentric heavenly spheres. These invisible spheres were not physical entities, but rather metaphysical ones— they defined the geometric structure of the heavens. In Pythagoras' cosmology the earth, along with the sun, moon, and planets, revolved around a body known as the central fire, a mysterious unseen entity that constituted a sort of purer, more ethereal sun. This unnecessarily complex addition to the celestial system was abolished in the third century B.C. by the great Pythagorean astronomer Aristarchus, who reduced the system to its logical simplicity by placing the *sun* itself at the center, with the earth and planets revolving around it. Here, then, was the heliocentric cosmology generally attributed to Copernicus, preempting him by 1,800 years. Fully two millennia before the invention of the telescope, the Pythagoreans had postulated what we today recognize as the structure of our solar system.

The mathematical harmonies inherent in the Pythagorean cosmos were the ratios between the sizes of the planetary orbits and between the speeds at which the planets move. Although ancient Pythagorean astronomers tried to determine these ratios, not till the age of modern astronomy could they be correctly calculated. The Pythagorean idea of the cosmos as a great mathematical harmony would be the inspiration behind the revolution in cosmology during the sixteenth and seventeenth centuries, for Copernicus, Kepler, and Newton were all consciously pursuing a Pythagorean quest. None of these men thought the heavens could be described in terms of simple ratios, yet all three believed the celestial realm could be described by mathematical relationships that they consciously regarded as harmonies of the cosmos. Indeed, as Einstein reminds us, the "longing to behold harmony" is *still* the driving force of cosmology.

The fact that Aristarchus' cosmology had to be rediscovered by Copernicus indicates that somewhere along the historic track this knowledge was lost. Although the Greeks were fascinated by Pythagoras, neither his science nor his philosophy ever acquired sufficient authority to become the dominant force in the ancient world.

While the Hellenes retained his belief that the heavens could be described by mathematics, they eschewed the heliocentric model, and the majority believed the earth was at the center of the cosmos, with the sun, moon, planets, and stars revolving around it. Likewise the Greeks eschewed the Pythagorean dream of a general mathematically based science, and retained math only for use in astronomy, optics, and a few simple cases such as Archimedes' study of levers. Thus the immensely promising start made in the sixth century B.C. was for the most part abandoned in the ancient world.

The chief orchestrator of the move away from heliocentric cosmology, and from mathematically based science, was none other than Aristotle, the most powerful of Greek philosophers. While Aristotle, like Pythagoras, was committed to explaining the world through rational means, he rejected mathematics as a suitable basis for doing so, since he did not believe it had the power to explain the *true* nature of things. Modern physicists have often derided Aristotle for his antimathematical stance and accused him of holding back the progress of physics for 2,000 years, but in fact Aristotle's rejection of math was quite justifiable, because it could not answer the questions *he* regarded as important. That the Greeks did not side with Pythagoras, but instead embraced Aristotle's approach to science, demonstrates that however much we may feel that mathematics is the right approach to physical science, it is not the obvious choice at all.

The fact that the Greeks had encountered the idea of mathematically based science and chose *not* to pursue this approach raises the immensely important question of why modern Europeans *did* eventually choose that option. The Greeks' conscious rejection of a mathematical science reveals that, like all other human activities, science too is driven by cultural choices. It is determined by what a society *wants* from its science, what a society decides it *needs* science to explain, and finally what a society decides to *accept* as a valid form of explanation. The Greeks turned away from Pythagoras and embraced Aristotle not because they were ignorant, but because Aristotle's science appeared to fulfill these functions better for *them*. A primary aim of this book is to examine how and why modern Europeans chose a Pythagorean approach. What eventually convinced people that a mathematically based science would be better for *us*?

But before we leave the ancient era, we must consider what became of Greek science in general, for, with the rise of Christianity, almost all of both Pythagorean and Aristotelian science was forgotten. For close to a millennium, the West lost so much contact with its ancient scientific heritage that virtually all of it had to be reacquired in the late Middle Ages. What happened to this extraordinary intellectual treasure? The astronomy, the mathematics, the Aristotelian physics and biology—what became of them? How did such a wealth of material simply disappear from Western consciousness?

The answer often presented is that Christians eradicated it. In this scenario, early Christian theologians are portrayed as purveyors of darkness who stamped out the glorious flame of Greek mathematics and science and plunged Europe into a "dark age" lasting a thousand years. While this view fits neatly with the idea that science and religion are inherent enemies, historical evidence does not support such an interpretation of events. In the past few decades, research has revealed that the reasons for the demise of ancient science are a good deal more complex. In particular, by the time of Christ, Greek science was *already* in a state of parlous decline. The "golden age" had ended centuries before, and, although science was still a subject of interest, few new ideas were emerging. Without a continuing stream of fresh ideas, the tree of ancient science was atrophying.

After Christianity began to be a serious force in the third century, some prominent Christians, such as Tertullian, opposed all "pagan" ideas, but others, such as Augustine, valued the Greek heritage and argued that Christians should utilize this knowledge for their own purposes. In particular, Augustine argued that Greek science could be used to help interpret biblical statements about nature. Far from being opposed to Greek learning, initially many Christians continued to educate their sons in the Greek tradition, but once the Christian community began to establish its own educational network, these new centers of learning focused on spiritual development rather than secular philosophy and science. In short, Christians did not stamp out Greek science; their interests simply lay elsewhere, and meanwhile, the culture that had given rise to this science had itself disintegrated. Greek science died in the West not through persecution but because there weren't enough enthusiastic practitioners to keep the tradition alive. Ultimately, the mantle of the Greeks passed to the Islamic world, where, in the bosom of Allah, the Hellenic heritage was kept in custody until Western interest rekindled.

The last burst of ancient Western science occurred in Alexandria in the closing centuries of the Roman Empire. While Rome was the political center of the empire, the city of the legendary library had long been its intellectual hub. The fate of this final phase of "Greek" science is symbolized by the story of its last great practitioner and our first major mathematical woman: Hypatia of Alexandria, who lived in the late fourth century. Although she was the earliest woman scientist whose life is well documented, none of Hypatia's work has survived, as with Pythagoras. Science historian Margaret Alic has noted that for 1,500 years she was considered the *only* woman scientist in history. Not until Marie Curie would the world at large know about another, although, as Alic and others have documented, many certainly existed.

Hypatia was born in an age deeply influenced by Aristotelian misogyny, when women were widely regarded as less than fully human. Nevertheless, she received a first-class education. In a scenario we shall see repeated many times throughout this book, her rare good fortune was due to the enlightened attitude of her father, Theon, a mathematician and astronomer who taught her himself. Indeed, up until the twentieth century, almost all mathematical women were taught by a male relative, usually a father or husband. Legend has it that Theon determined his daughter should become a perfect human being, and according to ancient sources she was—wise, learned, virtuous, and beautiful. Following in her father's footsteps, she too became a renowned teacher of mathematics and philosophy.

By the fourth century, the increasing pressure of religious sectarianism had resulted in Christians, Jews, and pagans attending

separate schools, but Hypatia taught everyone, and her home became a center where scholars gathered to discuss philosophical and
scientific questions. In addition to teaching, Hypatia wrote texts
on mathematics. As was common at the time, these texts were
generally in the form of commentaries on the works of earlier
mathematical giants, such as Euclid, Apollonius, and Diophantus.
In her books Hypatia included new solutions to old problems and
new problems to challenge astute students. She also contributed to
her father's mathematical and astronomical texts and compiled tables of the positions of celestial bodies. In addition to her theoretical work, Hypatia was interested in mechanics and practical
technology and designed several scientific instruments, including a
plane astrolabe used for calculating time and measuring the positions of the sun, stars, and planets.

Fourth-century Alexandria was not only the last ancient center
of Greek mathematics and science, but not unrelatedly, also a center of the Pythagorean revival that flourished in the late Roman
Empire. Hypatia was one of many intellectuals who belonged to
this movement, which is generally known as Neoplatonism. The
Neoplatonic era was the time, referred to earlier, when many latter-
day followers of the Samian sage promoted Pythagoreanism as a
Hellenistic alternative to the rising tide of Christianity, and touted
Pythagoras as a rational alternative to Christ. Both ancient commentators, Iamblichus and Porphyry, had written their books in
part with this intention. Thus Neoplatonism was perceived by
Christians as a religious rival, which, in truth, it was.

In 412 a fanatical Christian named Cyril became the patriarch
of Alexandria and began a campaign to rid the city of both Jews
and Neoplatonists. As a high-profile member of the latter group,
Hypatia attracted attention, but she refused to convert to Christianity. That commitment cost her her life. In 415, she was set
upon by a mob of Christian zealots, dragged from her carriage and
beaten to death. In the account of a fifth-century author: "They
stripped her stark naked: they raze the skin and rend the flesh of
her body with sharp shells, until the breath departed from her
body: they quarter her body: they bring her quarters unto a place
called Cinaron and burn them to ashes." In such a climate, Neo-

platonism could not survive, and with its demise the last phase of ancient science came to an end—a demise that Hypatia's murder has come to symbolize. The great era of Greek mathematical science, which began with the birth of a man, finished with the death of a woman.

2

God as Mathematician

THE INTELLECTUAL HISTORY OF THE CHRISTIAN WEST IS REALLY TWO histories: one of men and one of women, and at no time has the difference between the sexes' intellectual experience been so evident as in the High Middle Ages (roughly 1100 to 1400). While this period witnessed the revival of Greek mathematics and science, it also fostered the disenfranchisement of women from the culture of learning. Thus the High Middle Ages was a time of both triumph and shame. But in order to understand the Janus-like character of this remarkable age, we must begin somewhat earlier, not that long after the murder of Hypatia. There, in the so-called dark ages, we begin to see the central role the Church would play both in the revival of ancient knowledge and in the educational fate of women, for the exclusion of women from the later culture of learning was by no means a foregone conclusion. That things could have been different, that women too could have been part of the medieval "renaissance"—and therefore part of the revival of mathematically based science—is apparent in the opportunities that *were* open to them during the preceding centuries.

The first millennium of Christianity is often regarded as an era of unmitigated darkness for women, but while it is certainly easy to find examples of misogynist views among the patristic fathers, it is a mistake to infer from this that women of the early Middle Ages were utterly without power or voice. Indeed, during the first millennium, the Church was not the monolithic organization it would later become, and Christians in different countries found a wide variety of ways to express their faith. For example, early Christian men and women found ways to live and study *together*. In the seventh and eighth centuries, England, Spain, France, Italy, Germany, and Ireland all had traditions of double monasteries that housed both men and women. Although living quarters were segregated, the sexes shared schools and scriptoria and participated together in divine services. Some of these double houses were distinguished centers of learning, and not infrequently they were headed by women.

In Ireland, Saint Brigit established a double monastery at Kildare as early as the fifth century, and in England (where the tradition was especially strong), Ely Cathedral is on the site of another former double monastery, founded in the seventh century. Here there was a long line of abbesses, starting with its founder, Etheldreda, who was followed by her sister Sexburga, then by Sexburga's daughter Ermenhilda, then by Ermenhilda's daughter, Werburga. This monastic matriarchy demonstrates that early medieval Christianity was a good deal more flexible and complex than the typical view of the "dark ages" suggests. The fact that an abbess could pass on the mantle to her daughter implies a radically different model of religious community than anything endorsed by Christian churches today. Hilda, the grandniece of King Edwin of Northumbria, founded a double monastery at Whitby that became an educational center for the whole of England. At least five future bishops were educated there, and nuns as well as monks studied Greek and Latin. According to historian Suzanne Wemple, the double monasteries served as "co-educational schools," and in so doing continued a tradition that harked back to the study circles of the earliest Christian communities. Even more important than the double houses were the many nunneries of the early Middle Ages, which were a locus throughout Europe for the education of women.

In the late eighth century, however, a new element in the culture of Christianity began to threaten the educational opportunities available to women. Although women had not previously had equal opportunity with men, from now on the intellectual divide between the sexes would grow even wider. The first phase of this process took place during the reign of Charlemagne, emperor of the Carolingian Empire, which encompassed all of modern France, Belgium, Holland, and Switzerland, and parts of Germany, Austria, and Italy.

At the end of the eighth century, Charlemagne embarked on a campaign to reform the priesthood within his domain so that the Church could better serve as a spiritual wing of the state. Acutely, Charlemagne realized that a better clergy would have to be a better educated clergy, so he decreed that schools be set up at cathedrals and monasteries throughout his empire. This empirewide reform program precipitated the Carolingian Renaissance and heralded the revival of learning in the West. However, because the reforms were aimed at functionaries of the Church, the new schools were open only to boys and men training for that career. Girls and women could not attend these schools, and indeed, while new educational opportunities were opening up for men in the Church, female religious communities were placed under tight episcopal control, with women in these communities increasingly confined in their convents. As a consequence, the power and influence of women's religious houses began to decline.

The Carolingian Empire, and with it Charlemagne's reform movement, collapsed in the late ninth century, but it had set a powerful precedent. In the eleventh century, the papal see instituted its own reform movement, and this time, because the impetus came from within the Church, the reforms would be more enduring. One of the primary aims of these Gregorian reforms was to consolidate the power of the papal see and to make the pope the absolute authority over all Latin Christendom. He was to become, in effect, the de facto emperor of Europe, albeit a holy emperor. The reforming popes imposed a centralized and hierarchical structure onto the entire church system, taking power away from local bishops and vesting it instead in Rome. Monasteries were no longer allowed to operate with the autonomy they had formerly enjoyed and had to adopt a standardized code of behavior. Again, the reforms were used to strictly limit the autonomy and activities of religious women's communities.

Following the example of Charlemagne, Pope Gregory VII (after whom the reforms took their name) also realized that education was a key to reforming the clergy, and in 1078 he ordered that all bishops were to have "the arts of letters" taught in their churches. A century later, the Third and Fourth Lateran councils of 1179 and 1215 mandated that "every cathedral church shall maintain a master to give free instruction to clerics in the church and to needy scholars." These cathedral schools became leading educational centers of the twelfth and thirteenth centuries, and again, because they were set up to train church functionaries, women and girls could not attend. The effect of the Carolingian and Gregorian reforms was to alienate women from the mainstream of learning and education. Although a few women still had access to education in nunneries, the standard there was rarely on a par with that in the cathedral schools, and as time wore on the educational divide between the sexes widened, leaving women ever further behind.

One critical instrument of women's alienation from the educational mainstream was the adoption of Latin. Charlemagne had decreed that Latin was to be the official language of clerical learning, and boys spent years at cathedral schools studying Latin grammar and vocabulary. Because Latin was no longer a living language, the only way to master it was through formalized education, to which women did not have access. Indeed, says historian Walter Ong, with the use of Latin "a sharp distinction was set up between those who knew it and those who did not." It was the "secret language" of an elite club, which set the clergy apart from all other people. Most men were also excluded from this club, but because *no* women could become clerics, as an entire social group they remained on the outside. "In helping to maintain the closed male environment the psychological role of Latin should not be underestimated," Ong has asserted. It became, in effect, "a sex-linked language used only by males."

The declining role of women in the culture of learning during the late Middle Ages is apparent in the cases of the period's two most famous religious women: Hrosvitha of Gandersheim, a tenth-century nun, and Hildegard of Bingen, a twelfth-century abbess. Hrosvitha, a playwright and poet famous throughout the late Middle Ages, was one of the first outstanding female literary figures in Europe. Her writings also reveal a knowledge of the crude monastic mathematics of the day. By the standards of her time, Hrosvitha was a learned person, yet in the prefaces to her works she constantly apologized for herself, painfully revealing the prevailing attitude toward a woman who would presume to write: "I did not dare lay bare my impulse and intention to any of the wise by asking for advice, lest I be forbidden to write because of my clownishness. So in complete secrecy, as it were furtively . . . I tried as best I might to produce a text of even the slightest use." Elsewhere she justified her authorial audacity by declaring that her gift of writing came not from herself but directly from God.

Two centuries later the nunneries of Saxony also produced the most famous of all medieval religious women, Hildegard of Bingen (1098–1179). The youngest of ten children of a noble Rhineland family, Hildegard exhibited a religious inclination very early, and at the tender age of eight was sent to live at the Benedictine nunnery at Disibodenberg, where she was put under the tutelage of the Anchoress Jutta. In 1136, after Jutta's death, Hildegard succeeded her as head of the nunnery. During her long life, Hildegard was a prolific writer whose body of work includes two books on medicine and natural science, two books describing her visions, two biographies of saints, a book of exegesis of the Psalms, a play, and two books on a secret language she had invented. She was also a gifted painter and composer who left a legacy of extraordinarily beautiful liturgical songs and chants.

According to Hildegard's own accounts, the source of her tremendous creative output was the intense visions she experienced throughout her life. Although they began when she was five years old, she kept them secret until she was forty-two, only revealing them then because a powerful inner voice commanded her to do so and convinced her this was God's will. In describing the experience she wrote: "The heavens opened and a fiery light throwing off great streams of sparks utterly permeated my brain and ignited my heart and breast like a flame which does not burn but gives off heat the way the sun warms an object when it touches with its rays. And suddenly the meaning of the Scriptures . . . was revealed to me." Hildegard's claim of receiving visions from God was examined by a papal commission, and their authenticity was soon acknowledged by Pope Eugenius III. With this imprimatur she became a public figure, whose influence extended throughout Europe.

Yet, according to historian David Noble, her "remarkable reputation rested largely upon the evangelical standards of early Christianity . . . rather than upon the new intellectual criteria of the dawning High Middle Ages." For all her genius, this brilliant woman remained outside the new culture of learning emerging from the cathedral schools. She herself always claimed to be an uneducated woman, whose knowledge came not from formal learning but from her visions—that is, from God. As had Hrosvitha, she never claimed personal authorship, and she referred to herself as "God's little trumpet." Historian Gerda Lerner has noted that this tactic was adopted by many subsequent medieval women. Unable to speak with authority in their own voices, they justified their words by the authority of the Supreme Being. But, as Lerner has stressed, such a path was by no means free from risk. There was also danger in invoking the Lord—as was apparent in the case of Joan of Arc, another visionary, who was burnt at the stake as a heretic at the age of nineteen.

A further aspect of the Gregorian reforms that had a profound effect on women's access to education and learning was the reformers' determination to enforce chastity among the clergy. In contrast to the Roman Catholic Church today, during the first millennium of Christianity, many priests had wives or mistresses and fathered children. The idea of celibacy emerged from the monastic movement, which had initially been seen as the extreme end of Christianity. Toward the end of the first millennium, however, there were increasing moves by ecclesiastical authorities to force all clerics to be chaste. Until the Gregorian reforms these attempts had largely failed, and many priests continued to live openly with women and children, but when Pope Gregory VII came to power, he instituted strong measures. Married clergy were now defrocked and imprisoned. Gregory told noblemen they could defy married clergymen (who were no longer to be recognized as priests) and could seize their lands—an opportunity many nobles were happy to take. Some married priests were even killed. Not surprisingly, celibacy eventually became the norm. By contrast, priests in the Eastern Orthodox Church have always been allowed to marry.

At the heart of the Gregorian reformers' determination to impose celibacy was not so much hatred of women per se as a desire to accumulate and consolidate property in the hands of the Church. With no families, clergymen would have no children making claims on land or other valuables, and so it could remain under Church control. But one unfortunate by-product of this policy was an efflorescence of misogyny within the clergy, a sentiment that would spill over into academe.

The first universities were opened at the turn of the thirteenth century: Bologna around 1190, Paris around 1200, and Oxford around 1210. These new institutions of higher learning were outgrowths of the cathedral school system, and like those predecessors were set up to be training centers and recruitment grounds for male church functionaries. Again, women could not attend. This time, however, the consequences would be more far-reaching, because the universities were also the centers through which the heritage of the ancient Greeks was being revived. Because women could not go to university, they were effectively barred from participating in the revival of philosophy and mathematics that occurred in the High Middle Ages. Indeed, very few women participated in academe until the late nineteenth century. With few exceptions, not until then did women gain entry to these hallowed male sanctuaries, and many university departments did not admit women until the twentieth century.

Because universities were training grounds for the clergy, academics were also supposed to be celibate. Yet, ironically, as historian William Clark has noted, the rise of prostitution in Europe was intimately entwined with the evolution of the university. According to Clarke, "Prostitution first became generally tolerated in Europe in the twelfth century," and many brothels of the time were built close to universities, students being a significant clientele. Yet if the academic community did not entirely avoid women, the prevailing attitude among the ivory towers toward the female sex was far from positive, as is evident in the popularity of a famous medieval antimatrimonial treatise by Walter Map, archdeacon of Oxford. Map composed his treatise in the form of a letter to a friend to dissuade him from marrying. "Women," he wrote, "journey by widely different ways, but by whatever windings they may wander, and through however many trackless regions they may travel, there is only one outlet, one goal of all their trails, one crown and common ground of all their differences—wickedness." One might be able to visit a prostitute, but women as a rule were to be kept at arm's length. In 1290 a divorced scholar at Paris had to swear he would not be reconciled with his wife; if he did he would lose the right to teach, while Peter Abelard, the "father" of the University of Paris, was castrated by an angry mob when his marriage to Héloïse became public. Academic celibacy had a long and checkered history. Although in Italy we find married professors as early as the fifteenth century, in England, Fellows of Oxford and Cambridge universities were not allowed to marry until 1882.

Because medieval professors could not have families of their own (not within the legal bond of wedlock anyway), they too were officially childless. Thus any potential medieval Theon could not have had a daughter like Hypatia upon whom he could have lavished his knowledge. One outcome of the policy of academic celibacy was thereby to disenfranchise women from *indirect* access to academic learning as well. As far as Mathematical Woman was concerned, this was particularly devastating, because the universities were the *only* places where training in mathematics could be acquired. As feminist historians have now documented, women did in fact participate in the early development of most sciences, many of which had their roots in craft or home-based traditions in which women were active. Physics has been the exception because it did not derive from any craft tradition. In the early (medieval) years, it emerged almost solely out of the theocratic environment of academe; and even later, when it began to evolve outside the universities, practitioners still had to train there. Because mathematical training has been tied to academe, until women could crack this nut they were almost entirely excluded from participation in the field of physics.

One of the major purposes of the newly founded universities was to serve as centers for the study of the ancient texts that were coming back into circulation in the West. Between the tenth and twelfth centuries, most of the Greek work had been recovered—largely from the Arab world, where the knowledge had been kept alive by Islamic scholars. Thus, by the start of the thirteenth century, European scholars had reacquired much of the scientific corpus of the Greeks and Romans. Not surprisingly, this ancient influx sparked a flurry of new scientific activity—a mini–scientific revolution. During the thirteenth century, Albert Magnus wrote works on physics, astronomy, mineralogy, physiology, psychology, and medicine; Robert Grosseteste and Roger Bacon investigated lenses and mirrors and proposed new theories in optics. For proof of the vitality of thirteenth-century science, we need look no further than Petrus Peregrinus, who performed a sophisticated series of experiments that led to his discovery of the basic principles of magnetism, an achievement usually credited to the sixteenth century.

As it had been in the Greek world, the focus of thirteenth-century science was the work of Aristotle. Of all the ancients, Aristotle was by far the most prolific, and the sheer volume of not only his own work but also that of his many commentators assured that his comprehensive scientific worldview prevailed. Aristotle offered a complete cosmology as well as a complete biology and physics. No other scientist in history has covered so much territory, and, like the Greeks before them, the medievals fell under his persuasive spell. Indeed, Aristotelian thinking would dominate European science right up to the seventeenth century. An important aspect of the great logician's appeal to the medievals was that many aspects of his thinking—not the least of which was the teleological nature of his science—harmonized with Christianity.

In Aristotle's cosmology, everything in the world strove toward an ideal appropriate to its inherent nature: A stone dropped to the ground because it strove to return to its home, the earth; celestial bodies traveled in circles because the circle was the most "perfect" shape and thus the one most suited to their heavenly nature. Animate things strove to actualize the potential within themselves: The end goal of an acorn was an oak tree, and the acorn strove to actualize itself as the best possible oak tree. Similarly, the human soul strove to actualize its own potential perfection. This teleological worldview meshed well with Christianity, because God could be seen as the "force" that impelled things toward the perfection of their natures. Under Thomas Aquinas, Aristotelian philosophy was harmonized with Christian doctrine so that theology and natural science came together in a grand synthesis. Aristotle was Christianized and Christianity was Aristotelianized. Natural science could now be pursued within a Christian framework, and for the next four centuries it was.

Yet at the same time that Aristotelian science reigned, a Pythagorean strand of science, which aimed to describe the world in the language of mathematics, also reemerged. This development too was made possible by the recovery of the Greek corpus, for, along with the works of Aristotle, medieval Europeans also began to rediscover the works of the ancient mathematicians and astronomers. From the tenth century on there had been a special effort to reclaim astronomical texts, and along with them, the mathematical texts needed to facilitate the practical use of astronomy. This emphasis reflected the fact that an accurate understanding of the motions of celestial bodies was required to determine the religious calendar. The date of Easter, for example, is tied to the cycles of both sun and moon. Commensurate with this recovery of the Greek mathematical heritage, a Neoplatonic Pythagorean spirit also began to assert itself within the Christian subconscious. Once again philosophers were drawn to the idea that all is number. Paraphrasing Pythagoras, the twelfth-century Neoplatonist Thierry of

Chartres wrote that "the creation of number was the creation of things."

In fact, long before the twelfth century the ground was already laid for the reemergence of a Pythagorean spirit. Augustine himself, that greatest of all the patristic fathers, had been deeply influenced by Neoplatonism and had incorporated elements of it into his theology. Although during the first millennium of Christianity Western Europe had lost touch with mathematics, the Pythagorean idea of the universe as a great cosmic harmony had been tremendously influential throughout. Thus when mathematics was finally recovered, there was already fertile soil from which could spring a more full-blown expression of the Pythagorean spirit. In the late medieval context this took the form of reconceiving the biblical deity as a divine mathematician. In effect, Christians of a Pythagorean persuasion recast God as a mathematical Creator. As the thirteenth century theologian Robert Grosseteste expressed it: God "disposes everything in number, weight and measure." Grosseteste also referred to God as "the first Measurer." Here we witness the beginning of a process of not only Christianizing Pythagoras but also Pythagoreanizing Christianity. It would be from this Pythagorean Christianity that the science we know as physics would ultimately emerge. Half a millennium before the "scientific revolution," the seed of modern physics began to germinate in a theological revolution.

At the forefront of this movement was the aforementioned Robert Grosseteste (1168–1253), the first and perhaps the most profound of the medieval scientific thinkers. As a young man Grosseteste had studied at the cathedral school of Oxford, and after it was incorporated as Oxford University, he had been appointed its first chancellor, in 1215. Although one of the first scholars to write a Christian commentary on Aristotle's *Physics,* Grosseteste was also influenced by Neoplatonism and became the first medieval champion of mathematical science—initiating a tradition that made Oxford a leader in the field for the next two centuries. In 1221 he left the chancellorship, but his association with the university continued. He later became the first lecturer to the Franciscan order at Oxford, and, when he was made bishop of

Lincoln, he was finally in a position of considerable power over the academic community, for Oxford lay in the diocese of Lincoln and its schools were under the bishop's jurisdiction.

Today Grosseteste is remembered for his place in the history of science, but he saw himself first and foremost as a servant of the Church. In the early thirteenth century, the issues that had inspired the Gregorian reforms were still pertinent, and Grosseteste was one of the most zealous reformers of his day. In particular he was a tireless champion of chastity, and as both chancellor and bishop he insisted upon the strictest chastity among academics and religious men under his jurisdiction. All transgressions were punished severely. Nuns suspected of carnal infractions were dealt with particularly harshly. According to a contemporary account, the learned bishop regularly had his priests squeeze the breasts of the nuns to search for evidence of lactation, this being how Grosseteste believed "he could discover who was corrupt." Grosseteste also championed an ideal of life from which women and families were entirely absent; his image of the perfect learned community was an exclusively male world. And so while he created a glorious Pythagorean-Platonic metaphysics, as far as women were concerned Grosseteste sided with Aristotle—who had insisted that women were mentally defective and less than fully human. Women's supposed mental inadequacy was yet another ruse used in subsequent centuries to bar them from access to academe—the crucial locus of mathematical training.

From the beginning, Christian Mathematical Man's world picture was inextricably interwoven with his theology. In Grosseteste's metaphysics of light, we see the first full-blown expression of a mathematico-Christian cosmology, in which we may even recognize elements of the modern mathematical world picture. According to Grosseteste, the universe was generated from a point of primordial light—the divine illumination, or *lux,* of which visible light was said to be the physical manifestation. Now, because the definitive feature of light is that it propagates outward, like flares radiating from a candle, this original point immediately began to expand, forming the sphere of the universe. As the first emanation of God's power, Grosseteste believed that *lux* was ultimately the cause of all natural action in the universe. Indeed it was the primal force of the world. Man could not study the divine *lux* directly, but he could study its physical manifestation in light. Thus, Grosseteste believed light was the key to understanding the working of the natural world. Furthermore, because the Greeks had taught that light travels according to the laws of Euclidean geometry, Grosseteste concluded that a mathematical understanding of light would serve as the model for understanding all natural influence, or what we would now call force. As we shall see in a later chapter, this is close to what mathematical men believe today. In contemporary physicists' quest to understand the forces of nature, it is light that has generally served as the model.

Turning his metaphysics to practical effect, Grosseteste was one of the first medievals to advance the science of optics. A particular triumph was his insight into the formation of rainbows. Many ancient and Arab philosophers had considered this intriguing optical phenomenon, yet he was the first to propose that it was caused not by reflection, but by refraction of light—the same process that causes objects to look foreshortened in a pool of water. His suggestion inspired other scholars and culminated in the early fourteenth century with a sophisticated geometric explanation of the rainbow by Theodoric of Freiburg. From a theologically inspired metaphysics had thus emerged the beginnings of modern optics.

Of all medieval mathematical men, the most famous was Grosseteste's follower, the thirteenth-century Franciscan brother, Roger Bacon (1214–1292). Unlike Grosseteste, Bacon did not contribute anything significant to our mathematical understanding of the physical world; rather he distinguished himself by the force of his commitment to the new science. Throughout his long life, Bacon was a tireless campaigner for both mathematics and experimental science, and his three major works, all written for Pope Clement IV, were passionate attempts to convince the church hierarchy of the true value of mathematics and natural philosophy. Although many scholars were embracing Greek learning, in the thirteenth century some theologians were still resistant to this pagan incursion into Christianity, and Bacon appointed himself the chief defender against these naysayers.

In his campaign to sell science to the papacy, Bacon asserted that it would lead to all sorts of marvelous applications and inventions that would improve the human condition. Here he envisioned flying machines, automotive carriages, and machines for lifting heavy weights; ever-burning lamps, explosive powders, and a glass for concentrating sunlight to be used for burning enemy camps from afar. Following Grosseteste, he also envisioned optical instruments that would help the aged and weak-sighted to see, enable men to read small script at a great distance, and magnify objects so that one could count individual grains of sand. (Bacon and Grosseteste are credited with forecasting the telescope and microscope.) According to Bacon, science would also lead to improvements in agriculture and medicine and to elixirs for prolonging life.

Yet, in spite of these marvels, Bacon credited science first and foremost as "the handmaiden of theology." Contrary to a popular myth, he never pitted science against the Church, but went out of his way to enumerate the ways it could serve Christianity—sometimes going to bizarre lengths. For example, he suggested that science could be used to determine latitude and longitude more exactly, thereby helping to locate the ten tribes of Israel and perhaps even the Antichrist. He also suggested that optical devices could be used to terrorize unbelievers and that science could aid in the interpretation of Scripture, thereby proving articles of faith. Finally, science could be employed to make converts—an issue of considerable interest to the ecclesiastical hierarchy.

Ultimately it was in this respect that Bacon's influence may have been felt most strongly. He was convinced that if painters learned geometry and applied it to their craft, they could give images a literal verisimilitude that would make people truly believe in the events depicted. Miracles would seem real, he argued; the lives of Christ and the saints would become tangible and people would see for themselves the truth of the Scriptures. According to Bacon, God had created the world according to the principles of Euclidean geometry, and that was the way men ought to portray it. In his *Opus Majus,* Bacon extolled the virtues of solid-looking imagery and urged the Church to embrace the simulation of three-dimensional space in religious art. He called it "geometric figuring." By such imagery, he wrote, "the evil of the world would be destroyed by a deluge of grace."

The idea that geometric figuring could convert unbelievers and renew the fervor of the faithful soon took root. Art historian Samuel Edgerton has pointed out that just a decade after Bacon's appeal to Pope Clement, work began on the frescoes for the Franciscan order's new basilica in Assisi—suddenly a whole church was being filled with images deliberately painted to look as three-dimensional as possible. The life of Saint Francis was practically leaping out of the walls. For contemporary visitors these early attempts at perspective painting were as startling as the first photographs must have been in the nineteenth century, and the basilica at Assisi was soon the most visited church in all Western Christendom.

While painting had been slowly moving away from the flat style of early medieval art since the twelfth century, Bacon was one of the first to consciously champion the style we now know as geometric perspective. This shift was profoundly important for the history of physics, because with this new style of representation Europeans began to move toward a geometric understanding of space itself. Instead of paintings depicting spiritual or metaphysical hierarchies (with Christ being portrayed larger than angels, angels larger than humans, and so on), artists began to portray all characters equally in a single Euclidean space—Christ and angels became the same size as mortals. The ideal in painting was now to portray *physical* rather than *metaphysical* relationships, and to do so, painters had to become geometers—building on the work synthesized by the Greek mathematician Euclid around 300 B.C. Indeed, in the fifteenth century, the perspective painters were the leading mathematicians of the day. Their art was literally a sophisticated form of applied mathematics.

The geometrization of painting was not simply a matter of artistic taste, but emblematic of a huge shift in how Europeans conceived the world around them. Instead of focusing attention on spiritual relations, Western minds were increasingly turning their gaze to their physical environment. In turn, the evolution of perspective imagery became in itself an engine that further drove this fundamental shift in attention. Over a period of three hundred years, from the fourteenth through the sixteenth centuries, perspective painting trained European minds to think about space in geometric terms, and this profoundly contributed to the Western psyche's receptivity to a mathematical mode of thinking about motion. Both modern cosmology and modern dynamics were premised on a Euclidean notion of space—the same conception the perspective painters had been at such pains to understand. It is no coincidence that Copernicus developed his new view of the universe in the early sixteenth century, the peak of the perspective era. Nor is it a coincidence that Galileo (who taught perspective theory in an art academy) is the "father" of modern dynamics. Without the revolution in seeing wrought by Giotto and the geometer painters of the fourteenth and fifteenth centuries, there might not have been the revolution in science wrought by Galileo and the mathematical physicists of the seventeenth century. In the long run, the legacy of "geometric figuring" was greater than even Roger Bacon could have imagined.

The complexity of the task confronting those who first tried to develop a mathematical physics is apparent in the case of the fourteenth-century *calculatores,* an extraordinary group of mathematical men who followed in the path blazed by Grosseteste at Oxford. Their work in the study of motion came closer to the physics of the seventeenth century than anything in between, and Galileo himself was influenced by their achievements. Seminally, the *calculatores* developed a new conceptual framework for the study of motion; in particular they introduced the concepts of velocity and acceleration. While the idea of velocity might seem obvious to us today, as David Lindberg has pointed out, it is in fact an "abstract conception" which "had to be *invented* by natural philosophers." The invention of velocity highlights a major difficulty facing early physicists: They had to work out which particular aspects of the world around them could be usefully analyzed in mathematical terms. Velocity (the rate of change of distance over time) turned out to be a remarkably useful mathematical property, as did acceleration (the rate of change of velocity). But, lest we underestimate the difficulty in identifying such properties, we should note that the *calculatores* also attempted to apply mathematical analysis to qualities such as sin, charity, and grace. To them, these seemed as valid candidates for quantitative treatment as velocity and acceleration.

The *calculatores'* attempts to quantify spiritual as well as physical qualities are a potent demonstration of how much thinking changed between the fourteenth and the seventeenth centuries. Three hundred years after the *calculatores,* Galileo and Newton were successful with their physics precisely because they worked within a narrow range of purely physical properties. They contented themselves with trying to describe the motion of material bodies in Euclidean space and did not hanker after mathematical formulations of sin and grace as well. It was only when mathematical science became very narrowly focused that it really became successful. Yet if the attempts to quantify sin and grace seem quaint absurdities of a confused past, it is well to remember that we moderns have paid a high price for the narrow focus of modern mathematical science. The desire for a mathematics of grace speaks not of confusion but of a yearning for a world picture that encompasses both body *and* soul.

The fourteenth-century mathematical study of motion (epitomized by the *calculatores)* was in no small part inspired by religious considerations, at the heart of which was the issue of the limits of God's power. During the previous century, a debate had raged about whether God was compelled to adhere to the precepts of Aristotelian physics. Taking Aristotle's vision as their bible, a group of philosophers known as the Averroists had argued that God could not have created the world in any other way. Not surprisingly, traditionalists objected to the notion that the Lord was bound by Aristotle's imagination, and they declared that God could do anything he pleased, no matter how many Aristotelian tenets it might contravene. The matter culminated in 1277, when the bishop of Paris issued a series of decrees that asserted that

God's power was unlimited—that he could make *anything* happen, even if it outrightly *contradicted* Aristotle. Although this was essentially a theological issue, one by-product of the decrees was to free thinking about physics from the constraints of Aristotelianism. Released from these confines, mathematical men began their long quest toward a new science of motion.

It has recently been suggested by historians that one reason for the 1277 decrees and the general furor against the Averroists may have been their liberal attitude toward women. In opposition to Aristotle, the Arab philosopher Averroës held that men and women were essentially equal. Thus the very forces that set in train the flowering of fourteenth-century physics may also have been responsible for quashing a movement that could have led to more equitable treatment of women. Indeed the fourteenth-century Averroist Pierre Dubois did call for the education of women alongside men. Dubois had a vision of Christian expansion that required legions of missionaries, and in his mind women as well as men were to be recruited to the cause. Thus they too would have to be educated. In this deeply Aristotelian age, when many men regarded women as mentally defective, Dubois' stance was revolutionary. Needless to say, his ideas were not taken up, and the institutions of learning remained closed to women.

But as the High Middle Ages gave way to the first stirrings of Renaissance humanism, the tenor of Western culture began to change, and women themselves began to challenge the male hegemony of learning. Academic misogyny was the target of a feminist attack by Christine of Pisan (1364–1430) at the start of the fifteenth century. The books Christine read were, of course, all written by men, and she found them so filled with slander against women that, in her own words, she was left "despising of myself and all womankind." But she rallied from her depression and wrote her famous *Book of the City of Ladies,* a brilliant and heart-rending defense of women. Against the Aristotelians, Christine declared that the reason there had been few learned women in history was not that women were incapable of learning but rather that men had prevented them from rising above ignorance by denying them access to "the precious spring" of knowledge. With access to education, Christine believed that women also would begin to bloom: "If it were the custom to put the little maidens to school and . . . to learn the sciences as they do to the man children . . . they should learn as perfectly."

While Christine of Pisan was fighting for women simply to have access to education, our next mathematical man was receiving the finest Europe could provide. A model "Renaissance Man," Nicholas of Cusa grew up to be the premier fifteenth-century champion of mathematical science. According to the great Dutch historian of physics Eduard Dijksterhuis, Cusa's conclusions in this field were "so far-reaching [that] a revolution in thought would have resulted if they had been adopted and translated into practice." As it was, most of his ideas would not be put into effect until two centuries later—by which time they had been independently discovered by others. Yet while Cusa was one of the true forerunners of the "scientific revolution," as with Grosseteste and Bacon he too located his science firmly within a theological context. A cardinal of the Roman Catholic Church, Cusa was first and foremost a religious metaphysician who wove together mathematics and theology to create one of the most glorious syntheses of faith and reason the West has ever produced.

In true medieval tradition, God was both the starting point and end goal of Cusa's metaphysical speculations. For him, the universe was the unfolding of forms already enfolded within God. Accordingly, to know the world was to know the unfolding of God, and the way to knowing was through *number*. Cusa believed that number was nothing less than the "image" of "God's mind"—thus to study mathematics was to study the mind of God. Here then was the pinnacle of Pythagorean Christianity: God and mathematics harmonized into a mystical theology that combined both prescriptions for spiritual transcendence and scope for a genuine mathematical science of nature. In this respect Nicholas of Cusa was a kind of latter-day Pythagoras, a man for whom the spiritual and the mathematical were totally inseparable. For both men the primary purpose of mathematical study was to bring us ever closer to the undivided Oneness that is the source of all, in Cusa's words, that we may be "elevated in accordance with the powers of human intelligence" so we may come to see "the ever-blessed one and triune God."

Just as we were to know God through numbers, for Cusa we were also to know nature through numbers. In his treatise *The Layman*, Cusa declared that the way to knowledge about the world was through *measurement*, and he proposed that people who wished to understand nature had to engage in quantitative experimental research. Written in the form of a series of dialogues between a learned university philosopher and an uneducated layman, Cusa's text presents its case for quantitative science not through the word of the philosopher but through the layman. The philosopher, with his book learning, has come to hear the plain-speaking craftsman espouse his views on the world. The purpose of these dialogues was to demonstrate that true wisdom was to be found not in the ivory towers of academe but in the marketplaces and streets where ordinary men and women plied their trades and practiced their crafts. The growing respect during the Renaissance for the practical knowledge in crafts and trades was a major factor leading to the scientific revolution, and once again Cusa was at the forefront of this movement.

At the center of the layman's experimental universe was the humble scales. Using only this instrument, Cusa suggested an impressive range of experiments for investigating the physical world. He discussed the determination of density and the assaying of metal mixtures. He described how, from the increase in weight of a hydroscopic material such as a ball of wool, the humidity of the air could be determined. He explained how two periods of time could be compared by measuring the weight of water that flows out of a tank during each period. And he described how this method of timing could then be used to determine pulse rates and the speeds of ships. Furthermore, through weighing, even mathematical quantities could be ascertained. For instance, pi could be determined by comparing the weight of a circular dish of water with that of a square dish. Here at last was a clear conception of what was needed to realize a quantitative empirical science of nature. Here at last was a firm movement toward modern physics.

In Nicholas of Cusa we can see clearly the two great motivations that impelled Western minds toward an effective mathematical science: the theological and the practical. Supplementing the desire to behold God's cosmic mathematical plan were matters of immediate practical import. The need for an accurate calendar helped spur the development of astronomy, as did the need for reliable methods of navigation; the need for accurate maps spurred the development of geodesy; the desire for better cannons provided an impetus for the study of projectile motion; and the desire for machines provided a motivation for the development of mechanics and hydraulics. In allowing the theological strand to receive impetus from practical considerations, the modern West achieved a unique position with respect to mathematically based science. Historians have noted, for example, that, while the Greeks developed marvelous theories about the world, they never took the step of marrying theory to practical goals. It has been suggested that this was because they had a slave class and thus saw no need to develop labor-saving devices. Similarly, Joseph Needham, the great historian of Chinese science, has suggested that one reason the Chinese did not have a scientific revolution like the one experienced in Europe in the seventeenth century was that even though they were superb mathematicians, their philosophers remained too separate from their merchants. Again, they did not become sufficiently involved in solving practical problems.

By the end of Nicholas of Cusa's life, the Renaissance was in full flight. Leonardo and Botticelli were in their youth, while painting, architecture, sculpture, and engineering were flourishing in the warmth of a humanistic light. A new Age of Man was bursting forth from the long-legged boot of Italy. Central to this age was the reemergence of a classical Greek aesthetic, complete with its adoration of "perfect" proportion and mathematically inspired form. In paintings, buildings, and sculpture, artists and engineers took their inspiration from geometry and from the old Pythagorean idea of mathematical harmony. Leonardo urged: "Oh students, study mathematics, and do not build without a foundation." Both artists and philosophers were looking at the world around them in new ways. The time was set for a re-envisioning of the universe itself.

3

Ω μ δ Ω ω π Σ Δ γ β α ♄

Harmony of the Spheres

THE REENVISIONING OF THE UNIVERSE THAT WOULD BE MADE POSSI-
ble by the new mathematically based science was already presaged
in the works of the sixteenth-century painters, nowhere more
clearly than in Raphael's masterpiece the *Disputa,* one of the magnificent frescoes in the Raphael Rooms of the Vatican Palace. In
this image we witness the apotheosis of Roger Bacon's "geometric
figuring." The scene in the *Disputa* consists of two levels: In the
bottom half, a phalanx of bishops, popes, and saints are arrayed in a
semicircle on a marbled terrace. Above these earthly worthies floats
a matching semicircle of clouds, on which are seated Christ, the
Virgin Mary, and John the Baptist, flanked by the apostles. Behind
Christ's throne stands God, surrounded by a brace of flying angels.
The basic image has been painted thousands of times—it is the
classic Christian picture of heaven and earth. But what draws our
attention here is that instead of being portrayed as metaphysically
distinct spaces, the two realms have been united within a single

Euclidean space. Raphael's unification of heaven and earth not only marks the culmination of an artistic trend, it alerts us to the profound transformation under way in the European conception of the world around them.

At the same time that Raphael was working on his fresco, in Poland another Renaissance man was working on an image of heaven and earth, and he too was determined to portray the two realms according to the precepts of Euclidean geometry. This time, however, the image would not take the form of a dramatic tableau, but of a mathematical diagram showing the physical relationship between the earth and the celestial bodies. Perspective painting had so trained the Western mind to think about space in Euclidean terms that people were now extending that conception to the stars. In Nicolaus Copernicus' heliocentric image of the cosmos we have the ultimate perspective picture of the world.

If Copernicus can be seen as the culmination of Mathematical Man as Renaissance artist, he also is generally taken to represent the arrival of Mathematical Man as "modern" scientist. The first of the recognized "giants," Copernicus is the first person with whom contemporary physicists acknowledge a genuine kinship. But if users of radio telescopes are happy to include Copernicus as "one of us," it is no less true that he belongs in the same camp as our previous mathematical men, for he too was seeking God. Although his heliocentric cosmology was the first major triumph of modern mathematically based science, it also represents a milestone in the reconceiving of God as a divine mathematician. Indeed, the new cosmology of the sixteenth and seventeenth centuries was as much a triumph of Christo-Pythagorean inspiration as of mathematico-empirical progress. All three chief architects of this world picture—Copernicus, Kepler, and Newton—were profoundly religious men who forged their cosmic systems as offshoots of their theology.

Copernicus (1473–1543) came into the world at the height of the quattrocento, but his birthplace, the Polish town of Torun, was far away from the dramatic developments in Italy, and young Nicolaus grew up in an essentially medieval world. Yet in spite of this geographic impediment, his life was filled with opportunity. At the age of ten, his father died, leaving an upwardly mobile uncle,

Lucas Waczenrode, as the father figure in his and his brother's lives. A few years later, Uncle Lucas was appointed bishop of Varmia, and from this lofty position he used his influence to benefit both nephews. The first step was a good education, and Uncle Lucas sent both boys to the University of Cracow, which boasted excellent teachers of mathematics and astronomy. Nicolaus quickly displayed a passion for the latter, but at the time astronomy was not the basis for a career, and after Cracow, Uncle Lucas sent him to Italy to study canon law and medicine. Returning home with a doctorate in canon law, Copernicus had already secured a post as a secular canon at the Cathedral of Frauenburg, another favor organized by Uncle Lucas.

After the vibrancy of Italy, one cannot but wonder if Copernicus had reservations about returning to the Polish boondocks; he must have known that in Frauenburg he would find few people to talk with about the new ideas he had been exposed to in the great universities of Bologna and Padua. But if Frauenburg did not offer much in the way of intellectual stimulus, it did offer financial security—indeed, a veritable life of ease. Bishop Lucas had done well by his nephew, for canons at Frauenburg received a substantial income, as evidenced by the fact that they were expected to keep two servants in their employ. With plenty of leisure time, Canon Copernicus was ideally situated for a life of contemplation.

Copernicus' lifeline to the learned world was the printing press. Astronomers were one of the first groups to take advantage of the new technology of printing, and Copernicus was one of the first scholars to own a private collection of printed books. Some historians have suggested that without the printing press this isolated Polish canon could not have had a career in astronomy at all and that certainly without printing no one would have heard of him. The fruits of his labor became available only after his death, in his infamous book *On the Revolution of the Heavenly Spheres*. It is for the most part a monumental exercise in technical details, and very few people read it, yet few books in the history of science have been so important, not because it established heliocentric cosmology on a sound "scientific" basis—Copernicus did nothing of the sort—but rather because it demonstrated the mathematical validity of the essentially mystical belief that the sun was at the center of the universe. In terms of actual physics, Copernicus made no advance whatsoever. His real contribution was to faith—faith in an idea that in the hands of his successors *would* become the key to modern cosmology.

In the early sixteenth century, there was of course an existing cosmology—the one Europeans had inherited from Aristotle and his successors: The earth was at the center of the universe, and around it revolved the crystalline "heavenly spheres," which bore the sun, the moon, the planets, and the stars in great circles around us. Following Pythagoras, Aristotle had insisted that the circle and sphere were the only shapes perfect enough to describe the motions of the heavenly bodies. The problem was that even in Aristotle's own time it was known that the sun, moon and planets do *not* travel in perfect circles through the sky. The task of ancient astronomers had therefore been to show how the deviations from this perfection could be explained by combinations of different circular motions working in unison. In effect they had to explain the path of each heavenly body as the result of a number of celestial "gears." Think of a clockwork toy in which each individual gear turns in a circle but the combination of gears produces noncircular movements, such as the dance of a windup ballerina doll.

The pinnacle of this "clockwork" envisioning of the heavens was achieved by the Alexandrian mathematician and astronomer Claudius Ptolemy in the second century A.D. Ptolemy worked out a separate combination of circular gears for the sun and moon, and for each of the planets, which successfully accounted for each body's idiosyncratic motion through the heavens. Together they constituted a set of invisible celestial "mechanisms" from which astronomers could calculate eclipses, conjunctions, solstices, and other significant cosmic events. Ptolemy's system was also used for determination of the calendar. The accuracy of the system was not perfect, but given the tools available at the time it was an empirical masterpiece. The Arabs were so impressed they called his book the *Almagest,* which means, simply, "the greatest."

During the late Middle Ages, Europeans had reacquired this invaluable tome, and astronomers had been relying on Ptolemy's system ever since. Yet Copernicus was deeply offended by what he found in the *Almagest*. The crux of the problem, as he saw it, was not that Ptolemy wasn't accurate enough but rather that the ancient astronomer's system was clunky and disjointed, even downright ugly. As Copernicus said in the opening chapter of his own book, Ptolemy's cosmos lacked "symmetry" and "harmony." Furthermore, through a devious sleight of hand, Ptolemy had not adhered strictly to the sacred principle of circular motion—an unpardonable crime in Copernicus' eyes. Adamant that God had not created the universe as inharmoniously as Ptolemy portrayed, the canon of Frauenburg set out to find an alternative description of the heavens, one whose inherent "symmetry" and beauty would be worthy of "the best and most systematic Artisan of all."

In his quest for a more harmonious picture of the cosmos, a major issue Copernicus faced was that at the time astronomy was regarded merely as a mathematical game. Ptolemy's system of celestial circles was intended as a purely functional tool to provide a method for calculating celestial positions. It was never meant to describe cosmological *reality*. In the Aristotelian way of thinking, that task fell to philosophers, not astronomers. According to Aristotle, the role of astronomers was simply to improve their methods for calculating celestial positions and not to bother their heads trying to work out what was really going on up in the skies. Copernicus rebelled against this view and declared that the purpose of astronomy was to discover the true "structure of the universe." According to him, an astronomical system should be not just an aid to calculation but a genuine description of the actual cosmic plan. That plan was of course God's, and for Copernicus astronomy was a "divine rather than human science." Indeed, he declared, knowledge of the heavenly motions served "to draw man's mind away from vices and lead it towards better things," in particular, to "admiration for the Maker of everything." In short, for Copernicus, astronomy was a path to God himself.

Along with Nicholas of Cusa, Copernicus was an unabashed Pythagorean: His God was definitely a mathematical Creator—one who was also driven by Renaissance aesthetic ideals. For him, the heavens must *necessarily* exhibit aesthetic perfection. As to what constituted such perfection, Copernicus was in agreement with the perspective painters. To quote historian Fernand Hallyn: "If the Renaissance artist is often called a God, Copernicus' God create[d] like a Renaissance artist." In particular, the idea of symmetry was paramount to the artists of the time. As one Renaissance commentator explained: "Symmetry is the proper agreement between the members of the work itself, and relation between the different parts and the whole general scheme, . . . Thus in the human body there is a kind of symmetrical harmony between forearm, foot, palm, finger."

The idea of symmetry, and of the proper agreement between the parts, became central to Copernicus' vision of the cosmos, and echoing the Renaissance artists, he used the metaphor of the human body. He criticized Ptolemy's system because, as he said, it was as if "some one [had taken] from various places hands, feet, a head and other pieces and cobbled them all together." The result of this patchwork collage, according to Copernicus, was "a monster rather than a man." In Ptolemy's system each of the celestial bodies was described by its own set of circles, but there was no overall coherence—it was a disjointed set of individual "mechanisms." Copernicus believed the cosmos must be an organic whole in which all the parts were integrated, as in a properly proportioned and coherent body. Anything less, he believed, would be unworthy of the "best and most systematic Artisan."

The Renaissance artists' aim with perspective painting had been to portray accurately relationships between objects in three-dimensional Euclidean space. This was also Copernicus' aim with respect to the heavenly bodies. Unlike Ptolemy, who had merely been providing an aid to calculation, Copernicus intended his cosmic system to be a true image of the planets' paths through Euclidean space—a true perspective picture of the cosmos. But, as any artist well knew, one of the major tasks facing the painter was to decide from what perspective an image was to be portrayed. The eye of the artist could be positioned anywhere, but only from *certain* perspectives would the resulting image appear properly proportioned, and the true symmetry be revealed. This was the question Copernicus now faced: Where should the eye be positioned in order to see the full symmetry of the heavens? The answer he arrived at was that it should be positioned at the sun. From this perspective only, he declared, would the true harmony of the cosmos become clear.

By looking at the motions of the celestial bodies from the vantage point of the sun, rather than the earth, Copernicus found he could make a major simplification in the cosmic system—the trajectory of each planet could now be approximated by a single circle centered on the sun instead of several circles centered on the earth. Furthermore, the heliocentric system had an inherent "symmetry" because now there was a consistent *pattern* in the motions of the planets—the closer a planet was to the sun the faster it moved. Thus Mercury, the innermost planet, completes its orbit in 88 days; Venus, the next planet, takes 225 days; the earth takes 365 days; Mars, 687 days; Jupiter, 4,333 days; and finally Saturn takes 10,759 days (almost thirty years). In this scheme the earth fits naturally between Venus and Mars, and it was this beautiful internal coherence that so impressed Copernicus. As a realization of Renaissance aesthetic ideals, heliocentrism was a triumph.

Yet it was not the total triumph Copernicus had hoped for. That is because he ultimately found that in order to account for the idiosyncratic details of the planetary orbits, (the slight deviations from circular perfection), he had to add a number of extra little circles, just as Ptolemy had done. Following the Greeks, Copernicus accepted that circles were the most perfect shape, and hence the only ones permissible for celestial motion. Thus, he too believed it was his job to explain the orbits as some combination of circles. In the end then, his system became just as complex and cluttered as Ptolemy's. The sublime beauty of the grand scheme was tragically compromised. Furthermore, in the final analysis Copernicus' system was no more accurate than Ptolemy's. It wasn't any worse, but neither was it any better. This is an extremely important point, because if science is to be judged by the accuracy of its predictions—as modern scientists stress it should be—then Copernicus' system represented no scientific advance whatsoever over Ptolemy's. Its chief virtue lay in its aesthetic appeal.

But the problem with a cosmology based on aesthetics is that not everyone has the same tastes, and later astronomers weren't as entranced by Renaissance ideals. Indeed, as the sixteenth century progressed, those ideals became passé. Copernicus had painted a picture of a universe *he* thought to be worthy of God, but unfortunately not everyone agreed about the deity's proclivities. Echoing Copernicus' own criticism of Ptolemy, Tycho Brahe declared that Copernicus' system was itself monstrously proportioned, because all the planets were bunched together in the middle, leaving a huge empty space between them and the fixed stars. Brahe objected to the idea that God would have left a vast gap in the middle of the universe. Other astronomers had different objections, and so, as Owen Gingerich has shown, although Copernicus' system soon replaced Ptolemy's for the purpose of calculating celestial positions, almost no one in the sixteenth century accepted it as a real picture of the cosmos.

Aesthetic judgment was not of course the only thing impeding acceptance of heliocentrism. Since the dawn of time human beings have seen the cosmic order as a model for the human order—how people see the heavens reflects how they see themselves. For fifteen hundred years, geocentric cosmology, with its hierarchy of celestial spheres, had served as a model for the hierarchical order of medieval society. At the bottom of the celestial hierarchy was the lowly, corrupt, earth, seated in the middle of the universe. Moving away from the earth, one progressed through the consecutive layers of heavenly spheres, which got successively "higher" and more divine until, beyond the outer sphere of the stars, one reached the Empyrean Heaven of God. The degree of perfection of the spheres was in direct proportion to their proximity to God, and so the realm of humankind was the "lowest" because it was farthest away from him. Throughout the Christian era, this stratified cosmos had served as a justification for a stratified social order with peasants at the bottom and a king at the top.

But heliocentrism offered a radically different model. First, in this system the earth itself became a planet. As a full member of the celestial family, it could not easily be construed as the gutter of the universe. Second, following a tradition that originated with the Pythagorean worship of the sun-god Apollo, Copernicus placed

God at the sun—rather than beyond the distant stars. Thus, in his cosmology, the earth was not at the opposite end of the cosmos from God, but was in fact quite close to him. Copernicus also insisted that all the planets, including the earth, received the sun's light, and hence God's illumination, *directly*. According to him, humanity therefore experienced God's radiance unfiltered by any heavenly spheres. As a model for human social order, heliocentrism thereby undermined the justification for a strictly hierarchical society, and so was a threat not just to Ptolemy, but to a whole way of life.

The geocentric cosmos had also served as a model for the Church. The successive layers of celestial spheres symbolized the successive layers of ecclesiastics: priests, bishops, cardinals, and the pope at the top. In this model laymen, and all women, were at the bottom of the spiritual ladder, and their relationship with God had to be mediated by the intervening ranks of clerics. But heliocentrism as a model suggested that individuals could commune *directly* with the deity. Clearly this was a threatening proposition. The social and theological challenges symbolized by heliocentrism were in fact present in Renaissance culture long before Copernicus. As early as the fifteenth century, Neoplatonists such as Pico della Mirandola had been saying metaphorically that the sun was at the center of the heavens. Under the influence of Neoplatonic mysticism, a central sun was already a symbol that pervaded Renaissance imagery. But while Copernicus did not invent the sun-centered cosmos, Hallyn rightly points out that his cosmology was the empirical perfection of this idea. As such, it was the ultimate Renaissance symbol for a new world order.

Although astronomers did not immediately embrace heliocentrism, neither could they dismiss it, because Copernicus had made it an empirically defensible rival to geocentrism. If astronomy was a mathematical game, he had demonstrated that there were at least two viable solutions to the problem. Tycho Brahe soon proposed a third, a sort of hybrid between helio- and geocentrism. Thus, Copernicus had not resolved the question of the heavens; instead he had precipitated a crisis. Cosmologically, the late sixteenth century was an age of anxiety: What was God's real cosmic plan? Can we humans ever truly know it? And if so, how?

As the seventeenth century dawned, the beginning of the "modern" answer was being pieced together by one of the most enigmatic figures in the history of science. Although he is not nearly so famous as Copernicus or Newton, Johannes Kepler is the crucial link between them and is in every respect their equal. When Newton said, "If I have seen further it is by standing on the shoulders of giants," he referred to no one more than to Kepler. While Copernicus had confined himself to a game of celestial circles, Kepler created a genuine *physics* of the heavens. In doing so, he laid the foundation for Newton's law of gravity and the modern picture of the cosmos—thus becoming the first true mathematical *physicist*. At the same time, Kepler was one of the great mathematical *mystics* of all time. Following in the footsteps of "our true teachers, Plato and Pythagoras," he was led to his revolutionary cosmology by his own unique Christian Pythagoreanism. Of all mathematical men, none has been more ardent in his quest for cosmic "harmony" than this childhood runt who finally answered the great question of the age: How *do* the planets move through the heavens?

Johannes Kepler (1571–1630) was not the stuff of which legends are usually made. Myopic, neurotic, constantly suffering from real and imaginary diseases, and in his own words "doglike," he lived a life filled with hardship, financial struggle, and sickness. There was little in his miserable childhood to indicate the man he was to become, and nothing in his family history to hint at anything of the sort. Born in the provincial German township of Weil-der-Stadt, Kepler grew up, by his own account, surrounded by dissolute, vain, and vicious relatives. It was not an environment that fostered learning. His father was a mercenary adventurer who often left his wife and children to fight other people's wars and who moved the family around so much that, although Johannes was a precociously intelligent child, he could not attend school regularly. At the age of thirteen, however, he won a scholarship to a Lutheran seminary—his poor health and already avid interest in religion made a career in the Church the obvious choice. Of his youthful self Kepler once wrote: "He was religious to the point of superstition. As a boy of ten years when he first read Holy Scripture . . . he grieved that on account of the impurity of his life, the honor to be a prophet was denied him." In lieu of prophethood, he would settle for the role of humble clergyman.

Life at the seminary was more stable than at home—discipline was strict and classes began at four in the morning—but it was no less miserable. According to Kepler's biographer Arthur Koestler, "his fellows regarded him as an intolerable egghead and beat him up at every opportunity." At seventeen he entered the University of Tübingen, and after graduating from the Faculty of Arts he continued at the Theological Faculty in preparation for his chosen career. But just before his final exams, Kepler was offered a post as a teacher of mathematics and astronomy in Gratz, Austria. At university he had shown a great facility for these subjects, and when the senate was asked to propose a candidate for the job they recommended him. They may have been motivated to do so by the fact that Kepler had already shown himself to be an uncommonly quarrelsome young man with a tendency toward suspect theological views. Kepler eventually accepted this surprising offer because it meant financial independence—but only after ensuring that he could return at any time to finish his theological studies. He never did go back to Tübingen however, because one day while drawing a geometrical diagram on the blackboard for his students, he hit upon an idea that would propel him to the forefront of astronomy and finally lead him to a new cosmology.

At university, Kepler had learned about Copernicus' system and had immediately accepted heliocentrism as a real picture of the world: "I have attested it as true in my deepest soul," he later wrote. Nevertheless, he did not exhibit much interest in the subject until the day in Gratz when the figure on the blackboard suggested to him that he could explain the details of the heliocentric cosmos in terms of a beautiful underlying geometric pattern. Copernicus had discovered the general arrangement of the heavens— the sun at the center and the planets revolving around it. Now Kepler would explain precisely the orbital sizes and spacings. That there was a precise mathematical explanation for the cosmic plan was an article of faith with Kepler, because for him the world was a reflection of a supremely Pythagorean God.

Following Nicholas of Cusa, Kepler saw the world as the material embodiment of mathematical forms present within God before the act of Creation. "Why waste words?" he wrote. "Geometry existed before the Creation, is co-eternal with the mind of God, *is God himself.* . . . geometry provided God with a model for the Creation." Thus, "where matter is, there is geometry." Kepler saw geometrical form everywhere he looked: in the shapes of leaves and fish scales, in the forms of snowflakes, and in the spherical shape of the universe itself. Further, because he believed that the world was a reflection of God, who was a perfect being, according to Kepler it must *necessarily* be a perfect world, and therefore the manifestation of sublime geometric principles. "It is absolutely necessary," he wrote, "that the work of such a perfect Creator should be of the greatest beauty." It was precisely such underlying beauty that Kepler thought he had glimpsed in the figure on the blackboard in Gratz.

Just as geometry had provided God with the model for the Creation, Kepler believed that geometry was "implanted into man, together with God's own likeness." For Kepler, says Hallyn, "the human mind [was] a *simulacrum* of the divine mind," both being essentially geometrical. The implication was that man, as *mathematician,* was the true human reflection of God: that it was through mathematical study of the world that we could truly participate in the divine. Indeed, said Kepler, astronomers were "the priests of God, called to interpret the Book of Nature." In a letter to his old teacher he further declared: "For a long time I wanted to become a theologian. . . . Now, however, behold how through my effort God is being celebrated in astronomy."

The celebration of God was the whole purpose of the cosmological model that Kepler spun out of the diagram on the blackboard at Gratz. This figure gave him the idea that he could account for the details of the heliocentric cosmos by using the five Platonic solids—the only solids composed of perfectly regular faces. (One is the cube, whose faces are all squares; another is the tetrahedron, composed of equilateral triangles.) These five uniquely regular forms, discovered by the Pythagoreans and championed by Plato, had been the target of mystical speculation ever since. Kepler believed he could explain the precise sizes of the planetary orbits and the spacing between them by a configuration in which these five solids were nested inside a set of spheres—rather like a Russian matrushka doll. The cosmic plan would be this: the sun at the center enclosed by a small sphere, itself enclosed by one of the solids, then a larger sphere, then another solid, then another sphere, and so on. Six spheres would account for the orbits of the six planets, and the five solids would account for the spaces in between. The trick was to work out the correct order in which the solids should be arranged, so that the proportions of the model matched perfectly with the proportions of the real planetary system.

Kepler was not proposing that the sky was really filled with huge cubes and tetrahedrons; he was simply trying to discover the mathematical plan by which God had determined the precise dimensions of the cosmic system. He was looking for the underlying geometric beauty. With his scheme, the beauty consisted in the fact that it explained the entire cosmic plan, *spaces* as well as *orbits*, in terms of "perfect" geometric forms: the classic spheres, supplemented by the perfect polyhedra. Buttressed by a good deal of mystical prose, Kepler presented this idea in his first book, *The Cosmographic Secret*, one of the most peculiar and gushing texts in the history of modern physics.

Yet for all its aesthetic virtues, Kepler was not completely satisfied with his model because, once again, it failed to match the real planetary orbits *precisely*. He may have been one of the greatest mystics in the history of physics, but Kepler was also its first true empiricist. His desire to know the cosmic plan precisely drove him to demand an unprecedented level of accuracy. Errors that Copernicus had happily tolerated were unacceptable to Kepler, and he realized that in order to perfect his model he would have to calculate the orbits for himself, from the raw data. As luck would have it, he was living at just the time when a new set of highly accurate astronomical tables was being compiled. If only he could get his hands on this treasure, Kepler believed, he could unlock the secret of Creation.

The keeper of the jewel in question was none other than Tycho Brahe (1546–1601), and someone less like Kepler would have been hard to find. A giant of a man, with a huge stomach and a shiny metal alloy nose, his own having been cut off in a duel, Brahe had a lust for life and the money to indulge his passions. He loved food and wine and presiding at the head of a copious table; but above all, he had a passion for the stars and had dedicated his life to compiling the most accurate astronomical tables the world had ever seen. In doing so, he hoped to get data good enough finally to determine the true system of the cosmos. Rarely has history produced such perfect synchronicity: In Brahe and Kepler we witness the ideal union of the data gatherer and the theoretical interpreter.

The best observations require the best instruments, and Brahe had amassed these from all over Europe: sextants, quadrants, and armillary spheres. Backed by the king of Denmark, he had built a legendary castle-observatory, the Uraniborg, on an island near Copenhagen. Uraniborg boasted its own chemical laboratory, printing press, paper mill, intercom system, flush toilets, quarters for visiting researchers, and private jail, while the grounds had game reserves, artificial fishponds, and gardens. In the observatory stood a huge brass globe, onto which Brahe and his assistants etched the sky as they painstakingly remapped the positions of a thousand stars. But after twenty years lording it over his island fiefdom, Brahe left Denmark in a huff. He moved to Prague, where he had found a new patron in Emperor Rudolph II. There, in the early weeks of the year 1600, he and Kepler met. It was the dawning of a new century—the century when physics, as we know it, crystallized into its modern form.

Brahe guarded his data jealously, but in his heart he realized he was not the one who could unlock the secrets of this jewel, and to his credit, he recognized Kepler as one of the great mathematicians of his age. Brahe gave him the data for Mars, the planet that least fitted with either Ptolemy or Copernicus. Kepler was ecstatic and made a bet that he would work out its orbit in a week. Five years later he was still engaged in his "war on Mars." He tried countless circular orbits and combinations of multiple circles, just as Ptolemy and Copernicus had done, yet none of them matched the data with the accuracy he now demanded. At last he made a radical decision: He abandoned circles and decided that planetary orbits must be some other shape.

It is difficult today to conceive what a revolutionary step this was. For over two thousand years, *all* the great cosmological think- ers had proclaimed the circle the most perfect form, and hence the only shape suitable for the motions of heavenly bodies. Given that Kepler himself was an ardent Pythagorean, and fanatically devoted to cosmic perfection, it took an immense wrenching to free his mind from the circle dogma. Pythagoras and Plato might have adored the perfection of the circle, but Kepler realized that nature preferred something else. The question was, What? He had left the solid ground of tradition for terra incognita. Finally, after a great deal of thrashing about—of which he wrote, "I was almost driven to madness in considering and calculating this matter"—Kepler discovered that the shape of Mars' orbit was an *ellipse*. He later found this was true for all the planets.

Kepler's discovery of elliptical orbits heralded the emergence of modern cosmology because, instead of imposing a preconceived idea about the way the heavens *ought* to be, he had let himself discover the way they actually *were*. He had allowed the data to speak for themselves. Thus the ellipse represents the triumph of empiricism over dogmatism, of commitment to mathematical ac- curacy over submission to ancient authority.

Yet we would be wrong to think of this achievement as purely "scientific." For a start, virtually none of his scientific peers (in- cluding Galileo) accepted the aberrant form. That Kepler himself was able to accept a noncircular shape was not simply a reflection of his own commitment to empiricism, but also a reflection of the deep religious currents that permeated his thinking. If the "scien- tist" in him had discovered the ellipses, it was the theologian in him that provided the justification for their acceptance. And again, this theological impulse was profoundly Pythagorean.

The circle and ellipse are both members of a family of curves known as the conic sections. The circle was considered the perfect member of this family while the others were considered less perfect —they were all said to be "mixtures" of the circle and the straight line. Now, again following Nicholas of Cusa, Kepler associated the circle with the spiritual and the straight line with the material. The circle represented the Creator, the straight line the Created. (This symbolism is in fact widespread in both the East and the West.) Kepler had no doubt that ideally the heavenly bodies would travel in perfect circles, but he reasoned that, because they resided in the material world, their spiritual purity was diluted by the influence of matter, and hence the ideal circularity of their orbits was diluted with straightness—thereby rendering them ellipses. But even though the perfection God intended for them could not be fully realized in the material world, Kepler pointed out that because ellipses are mathematically very close to circles, the planets were *striving* to reach the divine ideal. In effect, they achieved the highest degree of geometric perfection that *nature* would allow. Thus Kepler, who began his astronomical career out of a fanatical obsession with perfect forms, introduced "imperfection" into the cosmic picture—and he justified doing so by an essentially religious argument.

With his ellipses, Kepler was the first to propose a mathematical picture of the cosmos that did not rely on invisible celestial "gears" churning behind the scenes. In his cosmos were just the sun and six planets tracing out their elegant curves. All the complicated circle machinery that had plagued Copernicus as well as Ptolemy had been jettisoned, and in its place stood a truly simple and harmonious system. Furthermore, Kepler insisted that planetary motions were the result of a real *physical* force. In Copernicus' system, the sun was at the center of the planets, but it played no role in causing their motion; God was responsible for that. Kepler, however, insisted that the sun emanated a force that caused the planets to revolve around it. Whereas Copernicus' sun was an inert *symbolic* center, Kepler's was an active *physical* center. Here then was the seed of the modern idea of gravity that Newton would codify later in the century. While Copernicus had placed the sun at the center of the planets, it was Kepler who transformed the heliocentric cosmos into a real physical system, thereby making him the first *astrophysicist*. Yet lest we think he was merely being a good "scientist," again he had a complex theological justification for his gravitational force. As always with Kepler, science and religion were inextricably entwined.

To the end Kepler was motivated by Christo-Pythagorean inspiration. His last major work was a rapturous dissertation on the ancient Pythagorean subject of cosmic harmony, in which he presented over a dozen mathematical relationships he had discovered in the motions of the planets. One of these would prove crucial to Newton in his search for the law of gravity. Today physicists single out just three of Kepler's "harmonies" and call them the laws of planetary motion, but for him they were all keys to God's cosmic symphony. Although he went to his grave with little acclaim, few mathematical men have been so rewarded in their lives: Like Pythagoras before him, Johannes Kepler died believing he had heard the divine harmony of the spheres.

Just as Mathematical Man was increasingly turning his gaze upward to the stars, so too it was in the field of astronomy that we begin to see the reemergence of Mathematical Woman. Because universities were still closed to them, women could not get access to the formal mathematical training needed to become great system builders like Copernicus and Kepler. Nonetheless, from the early seventeenth century, women found ways to participate in the grand new adventure of exploring the heavens above. They were able to do so because much of the new astronomy was taking place not within the universities themselves, but in private observatories, such as Brahe's Uraniborg. Outside the institutional setting of academe there were no formal rules to prevent women from participating—and so they did.

For instance, Tycho's sister, Sofie Brahe, often helped her brother in his observatory. Without higher education, women astronomers did not have the training to be theoreticians; rather, like Tycho himself, they became observers, star catalogers and astronomical table makers, meticulously watching the heavens and recording the dance of the celestial bodies in its myriad subtle complexities. But if this is not the work which is usually credited in history books, it is the lifeblood of astronomy—after all, it was Brahe's observations that had provided the basis for Kepler's theoretical revolution.

In Kepler's homeland, Germany, there was a particularly high concentration of early women astronomers. One of the first was Maria Cunitz (1610–1664), sometimes known as the second Hypatia. As was her Alexandrian counterpart, Maria was educated by an enlightened father, Heinrich Cunitz, a Silesian physician and landowner who taught her history, medicine, mathematics and languages, including the all-important Latin. Her main interest, however, was astronomy, another love she learned at her father's side. After marrying an amateur astronomer, Cunitz began work on a set of astronomical tables for calculating the positions of planets. Her main purpose was to simplify Kepler's own monumental but difficult planetary tables. She published her tables in the book *Urania Propitia* (1650), a text in which she also dealt with the art and theory of astronomy—introducing her readers to these technical subjects in a comprehensible fashion. However, the notion of a woman writing a book about a mathematical science was so novel that few people believed it was her own work, and in later editions a preface had to be added in which her husband asserted that he had taken no part in the effort. Whereas women authors of the late Middle Ages had to justify their audacity by appealing to the authority of God, Maria Cunitz now found that a mathematical woman who dared to write constantly had to defend her work as her own.

Cunitz's entrée to astronomy had come through her father, and, in fact, all the female astronomers of this era depended on the enlightened support of male relatives or spouses. For Elisabetha Koopman (1647–1693), the man in question was her husband, the astronomer Johannes Hevelius, whom she married when she was just sixteen years old, and he fifty-two. For ten years Koopman worked as her husband's assistant in their private observatory, making observations for a new star catalog, and after he died she carried on the work alone, eventually publishing the largest star catalog to date. Maria Eimmart (1676–1707) of Nuremberg was also trained by her father, an amateur astronomer and successful painter and engraver, from whom she learned art as well as science. Utilizing both skills, Eimmart became expert at making accurate astronomical drawings, which in this age before photography served a vital function for the scientific community. In the 1690s she produced 250 detailed drawings of the phases of the moon that laid the groundwork for a new lunar map.

Of all the female astronomers of the seventeenth century, the most outstanding was Maria Winkelmann (1670–1720). Yet it was precisely Winkelmann's success that would be her undoing. Again, as with Hypatia and Cunitz, Winkelmann was educated by her father; but in addition to this familial support she received advanced training from a local astronomer. At his house she met the man who would become her husband, Germany's leading astronomer, Gottfried Kirch. Winkelmann chose Kirch, a man thirty years older than herself, because she recognized that with him she could continue to pursue her passion for astronomy. At a time when women had no access to scientific equipment on their own, the only way to gain such access was through a man. And she could hardly have selected a more suitable candidate than Kirch. After they were married, he was appointed to the prestigious position of astronomer to the Berlin Academy of Sciences, and for the next decade, until his death, she worked at his side. Throughout the night, husband and wife took turns observing the heavens; while one slept, the other peered through the telescope.

One night during an otherwise routine stint of observation, Winkelmann spotted something that was anything but routine—a new comet. Comets may seem rather mundane today, but in the early eighteenth century they were important news. Waking her husband, Winkelmann alerted him to her discovery and he immediately sent word to the king; but since the report bore Kirch's name, as academy astronomer, everyone assumed the comet was his discovery. Privately, Kirch acknowledged that his wife had the priority, and when the report was reprinted some years later he formally acknowledged the finding as hers. If Maria Winkelmann had been a man, the discovery of a new comet would have ensured her a firm reputation in the astronomical community, and it would have guaranteed her a professional position of her own. Her husband's career was in part based on his discovery of a comet some twenty years earlier, and Tycho Brahe's reputation had been greatly enhanced by his discovery of one in 1577. Winkelmann was denied due credit for her discovery because she was a woman.

The degree to which the astronomical community was resistant to women became all too evident when Kirch died in 1710 and the position of academy astronomer became vacant. Despite the fact that Winkelmann had worked side by side with her husband, performing all the functions he performed, she did not presume to ask for the position of full astronomer. Rather she petitioned the academy to appoint her and her son as assistant astronomers in charge of calendar making. One of the academy astronomer's primary responsibilities was to produce the official calendar of the German lands, and in the later years of Kirch's life, Winkelmann had taken charge of this task. Although she had been doing the work in question for more than a decade, the academy council rejected her. Nobody questioned her qualifications, the issue was her sex. Academy secretary Johann Jablonski set the tone of the ensuing debate when he wrote: "Already during her husband's lifetime the society was burdened with ridicule because its calendar was prepared by a woman. If she were now to be kept on in such a capacity, mouths would gape even wider." A woman, then, even a manifestly capable one, was in her very being seen as a threat to the prestige of the fledgling academy.

One of the few people who supported her was academy president Gottfried Leibniz, co-discoverer (with Newton) of calculus, and one of the great physicists of the seventeenth century. In spite of Leibniz's support, the academy refused to accept Winkelmann, and appointed instead an inexperienced man who soon proved incompetent. Ironically, after his death, the position of astronomer was given to Winkelmann's son, and once again she worked as an unofficial and unpaid assistant for the academy. By this time, however, hostility toward her had become intense, and, because she refused to stay in the background when visitors came to the observatory, members of the academy soon demanded she leave the facility altogether. Thus, in the last years of her life, this first-rate astronomer was forced to hide herself at home. Without access to equipment, her astronomical career came to an end.

Maria Winkelmann's fate is symbolic of the collective fate of women in astronomy. Despite their lack of access to higher education, women quickly found ways to participate in the new science, but, as the field institutionalized, they found themselves barred from official positions, and from formal participation in the community. As long as women astronomers stayed at home, assisting male relatives, their presence was tolerated, and even welcomed— for they could be used to perform the meticulous, but routine, calculating work that is essential to much of astronomy. Indeed, throughout the eighteenth and early nineteenth centuries, there was a long list of distinguished "amateur" women astronomers. Notable among them were Caroline Herschel (1750–1848), who, working with her brother William, helped to found sidereal astronomy; and Nicole de la Brière Lepaute (1723–1788), who, with the mathematician Alexis Clairaut, first calculated the return of Halley's Comet. Although Clairaut originally gave Lepaute full credit for her contribution, he later retracted his acknowledgment, and today he alone is generally given the credit.

The fact that women of the seventeenth and eighteenth centuries *did* participate in the new astronomy is evidence that women *were* eager to be part of the "scientific revolution." Yet instead of welcoming them into the enterprise, many mathematical men tended to put up obstacles. If the norms of society at the time undoubtedly made it difficult for women to take up mathematically based science, its practitioners far too often made it even harder.

4

$$\# \quad \varnothing \quad \div \quad \leqq \quad \leqslant \quad \infty \quad = \quad + \quad \pm \quad \sim \quad \sqrt{} \quad >$$

The Triumph of Mechanism

IN THE INTRODUCTION TO HIS INFAMOUS BOOK, COPERNICUS CITED A number of ancient authorities to support his case for a new cosmology. Along with the Pythagoreans Philolaus and Heracleides, he invoked the Egyptian sage Hermes Trismegistus, "the thrice great." Trismegistus is not someone any scientist today would acknowledge as a forebear, yet during the Renaissance this shadowy Egyptian enjoyed an immense status. To many Renaissance minds, he was one of the foremost ancient philosophers and Copernicus' appeal to his authority was an attempt to add much-needed credibility to his own contentious case. Yet, unlike the Pythagoreans, Hermes Trismegistus was not a mathematician or astronomer but a magician; in fact, he was considered the source of Western magic. Why did the "founder" of modern cosmology appeal to the authority of a magician? What credibility could a practitioner of the occult arts offer a mathematical astronomer?

Although we in the twentieth century tend to dismiss magic as superstitious hokum, during the fifteenth and sixteenth centuries it was considered to be a legitimate science of nature. Along with the revival of ancient aesthetics, the Renaissance witnessed a revival of ancient magic, which many people regarded as a path to true and useful knowledge of the world around them. While the occult arts may have been viewed with suspicion by those who feared their potential for evil, they were certainly not dismissed. To Copernicus and his contemporaries, magic was a real power based on legitimate and subtle understanding of the hidden forces of nature. It is not known today whether Copernicus himself was a follower of Trismegistus, or whether he only invoked the name of the ancient magus in order to give his own ideas a wider audience. What is certain is that, at the time, the name of Trismegistus held wide appeal.

By the early seventeenth century, however, when Kepler announced his discovery of planetary ellipses, most mathematical men regarded Trismegistus not as a higher authority, but as an unwelcome competitor. Mathematical scientists and magicians were now rivals, with each camp claiming the true path to knowledge. Both believed that a new philosophy of nature was needed to replace the old Aristotelianism, but they agreed on little else. Although it seems obvious to physicists today that Kepler's planetary laws provided a powerful argument for the efficacy of mathematically based science, to his contemporaries that was far from clear. While it was widely acknowledged that mathematics was useful for describing the celestial realm, few people believed it could provide an understanding of the terrestrial realm. Magic, on the other hand, seemed to offer genuine insight into a broad spectrum of nature.

The battle to establish mathematically based science as the successor to Aristotelian science is one of the great sagas of modern Western civilization. At stake was the whole conception of nature and humanity's relationship to it. Intertwined with this were fundamental issues of power. Who was to have the epistemological power to say what was legitimate knowledge of nature and what was not? Who, in short, were to be the keepers of the flame of "truth" about the natural world? To understand the rise of "physics" in the seventeenth century, we must understand its practitioners' battle against magic, for in the heat of this battle the character of modern Mathematical Man was forged. It was in opposition to magic, and its heretical religious implications, that the new physicists created their philosophy of nature, one they carefully tailored to be compatible with orthodox theology. Despite the common belief that the seventeenth century marked a definitive break between science and religion, the new physicists wanted nothing so much as a philosophy of nature that harmonized with mainstream Christianity, the better to combat the socio-religious threat of magic.

The difference between "science" and "magic" was not always so clear as it seems to us today. As long ago as the thirteenth century, scientists such as Roger Bacon and Albert Magnus had been students of the occult arts. During the Middle Ages, however, one could not practice magic openly, because the Church disapproved of meddling with occult forces. But magic's acquisition of a new legitimacy during the Renaissance made it, for a time, the leading competitor to Aristotelian science. Furthermore, unlike the Aristotelians, the magi sought to apply their knowledge to a broad range of practical problems. Whereas the Aristotelians stayed cloistered in the ivory towers of academe, the magi went out into society and put their craft to use. In particular, they were employed at Renaissance courts.

The birth of Renaissance magic can be traced to the Italian priest and physician Marsilio Ficino (1433–1499), who worked at the court of the grand duke Cosimo di Medici (the Elder). Along with many Renaissance princes, Cosimo collected ancient manuscripts, and translating the Greek texts was among Ficino's tasks. Around 1460 one of Cosimo's scouts brought from Macedonia a document that immediately piqued the grand duke's interest, and he ordered Ficino to translate this treasure immediately. The Macedonian manuscript turned out to contain an account of an ancient Egyptian magical religion. Now known as the *Corpus Hermeticum*, this document was one of the key texts on ancient magic to resurface during the fifteenth and sixteenth centuries. On these so-called hermetic texts was built the complex and peculiar edifice of Renaissance magic.

The recovery of the hermetic texts was the last phase of Europe's campaign to reclaim the lost wisdom of the ancients. This particular knowledge, however, was thought to predate even the Greeks, for Renaissance scholars believed these magical texts were the writings of the first *Egyptian* sage—Hermes Trismegistus. The supposed antiquity of the texts lent them an extraordinary power in the fifteenth and sixteenth centuries because, with these documents, Ficino and his successors believed they were recovering the very font of Western knowledge. They saw Trismegistus not only as the original source of Egyptian wisdom, but also as the ultimate source of the ideas of the Greek giants Pythagoras and Plato. Here, they thought, was the *prisca philosophia*—the first and hence the purest knowledge of the world—a basis for a true understanding of nature and its hidden powers. For a century and a half, the hermetic texts thus provided a foundation for a philosophy of nature that challenged the long reign of Aristotle.

But in the early seventeenth century, careful philological study revealed that these magical works were not ancient Egyptian in origin, but rather had been written in the latter years of the Roman Empire. The Pythagorean and Platonic overtones they contained were not indicative of some prescient Egyptian sage but simply reflected the Neoplatonic atmosphere of late antique Rome. What is extraordinary, in retrospect, is how easily people accepted these documents as genuinely Egyptian. Yet that belief was not just a fabrication of the Renaissance; it had been simmering since the fourth century A.D. No less an ancient authority than Augustine had acknowledged Hermes Trismegistus as an Egyptian sage. By the time Ficino translated the *Corpus Hermeticum*, the myth was a thousand years old, and neither he nor anyone else saw reason to question it.

Echoing Neoplatonic thought, the hermetic texts espoused an *organic* philosophy of nature. The world was seen as a living organism with a material body and an immaterial "world soul." Beyond the material realm was a divine "intellect," the source of the "ideas" (or forms) that became manifest in nature. The world soul was the intermediary between this divine intellect and the material realm, and was itself associated with the *celestial* realm. The basic principle of hermetic magic was that the magus could manipulate the power of the world soul by drawing down the influences of the stars, thereby affecting events on earth and the affairs of humankind. To use the phrase of Renaissance historian Frances Yates, hermetic magic was "astral magic." The idea that the stars influence the terrestrial realm was not of course unique to hermeticism, for it was also the basis of astrology. Yet astrologers merely interpreted the stars and divined celestial portents; they did not attempt to manipulate or alter the astral influences. Practitioners of astral magic purported to do just that, thereby escaping the determinism inherent in astrology. Behind astral magic lay the belief that the link between the celestial and material realms could be made to work in both directions: The stars could influence the affairs of humans, and humans, in turn, could influence the stars. The task of the magus was to learn how to bend the astral powers to his command.

Practically speaking, each of the recognized celestial objects—the sun, the moon, the planets, and the twelve constellations of the zodiac—exercised specific influences that the magus had to learn to control. For instance, Saturn was the planet associated with abstract study and was supposedly inclined to influence melancholia. The sun and Jupiter were the seats of healthful influences, as was Venus, the planet of love. Each celestial body or zodiacal sign was associated with particular stones, metals, plants, and animals in which its influences could be concentrated. By manipulation of these objects, the magus aimed to enhance the desired astral influences and minimize the undesired ones.

A further aspect of astral magic was the use of talismans. In addition to stones and plants, each celestial body and zodiac sign was associated with certain images. By inscribing the right image on the right material at the right time and in the right frame of mind, the magus believed he could gain great power over the astral influences. One sun image consisted of "a king sitting on a throne, with a crown on his head and beneath his feet the figure (or magic character) of the sun." A Saturn image might be "a man with a crow's face and foot, sitting on a throne, having in his right hand a spear and in his left a lance or an arrow"; and for Venus, "a woman with her hair unbound riding on a stag, having in her right hand an apple, and in her left, flowers, and dressed in white garments." The making of talismans was a complex and difficult art that required a thorough knowledge of astronomy, mathematics, music, and metaphysics. According to the hermetic texts, no one could expect to succeed in this intricate art unless he was a thorough student of philosophy, and Renaissance magic was, therefore, a highly intellectual pursuit.

What could be achieved with this intellectual magic? Among other things, the hermetic texts listed procedures and talismans for succeeding in business, escaping from prison, overcoming enemies, attracting the love of a desired person, curing diseases, and promoting health and longevity. The medical use of magic was what had particularly attracted Ficino. Again, the idea that the stars affect health was by no means new, but when Ficino published *The Book of Life*, he presented a new and subtle use of talismanic medicinal magic. The book was intended primarily for students, whose intense study inclined them to illness and melancholy. Their occupation brought them under the influence of Saturn, a planet inimical to the vital forces of life and youth. Ficino advised melancholy students and the old (whose life forces were declining as a matter of course) to avoid stones, plants, and animals associated with Saturn, and to surround themselves with those associated with the sun, Jupiter, and Venus. Such people were advised to wear ornaments of gold (a solar metal full of jovial spirits), and because green was a Venus color, they were to take walks in the country, there to pick solar and jovial flowers, such as roses and crocuses. They were also instructed about a non-Saturnian diet and told to inhale pleasant scents. Finally, Ficino advised them to make talismans: To obtain a long life one was to engrave an image of Jupiter on a clear white stone, or an image of Saturn on a sapphire. Clearly this was not medicine for the poor.

As a priest, Ficino was well aware that by advocating the use of magic he was treading on explosive theological territory, so he was at pains to point out that his magic was not demonic, but only evoked *natural forces*. It was based, he claimed, on a deep understanding of the ways of nature. Fearing to overstep theological boundaries, Ficino did not venture to apply magic beyond the domain of health, and made only modest claims for its power, but many of those who followed had much grander aims. His younger contemporary Pico della Mirandola went far beyond Ficino's claim of natural magic by introducing into the hermetic tradition the magic of the Jewish Cabala. According to Mirandola, cabalistic magic was much more powerful than purely astral magic because it enabled its practitioners to draw down not only the natural influences of the stars, but also the supernatural influences from the realm of the divine intellect. Mirandola's was not just star magic but angel magic, which supposedly involved a deep understanding of the names of God. Cabalistic magic also involved a numerology based on the letters of the Hebrew alphabet and on biblical texts. Yet it was precisely this mixing of magic with religion that would ultimately prove detrimental to its cause.

At the same time that the magi aimed at a deep understanding of nature, theirs was also a profoundly religious movement, for as the source of divine knowledge, the hermetic texts were also viewed as a complement to the Scriptures. Indeed, because one hermetic text contained an account of Genesis that bore a striking similarity to the biblical version, Renaissance hermeticists regarded Hermes Trismegistus as nothing less than an Egyptian Moses. In addition, several other texts appeared to prophesy the coming of Christ. The hermeticists interpreted these passages as evidence that Trismegistus was not only the precursor to the Greek philosophers but also the first prophet of Christianity itself. To them he was the original link connecting *all* the great strands of Western wisdom, religious as well as secular. It was this supposed original synthesis they sought to restore by integrating hermeticism back into Christianity. Where the fusion of Aristotelianism with Christianity had defined late medieval culture, the Renaissance magi sought to define post-medieval culture by a fusion of Christianity and hermeticism.

The religious threads of hermeticism were evident from the

beginning. Countering the Church's antipathy to magic, Ficino justified it on the ground that the magus operated from a deep understanding of the process by which the Ideas in the divine intellect become embodied in matter. In effect, he said, the magus was a semi-divine operator who utilized the same methods by which God had created the universe. As one historian has noted, the practice of magic was viewed as a "holy quest," a "search for knowledge, not by study and research but by revelation." "The art of magic is the art of worshiping God," wrote Sir Walter Raleigh, while Mirandola defended the hermetic texts by declaring that they demonstrated the dignity of Christ. Visible proof of the religious undercurrents in Renaissance hermeticism can be seen in the cathedral of Siena, where the first thing that greets the visitor upon entering is an image of Trismegistus inlaid into the mosaic pavement. This place of honor was bestowed in recognition of his supposed role as the first prophet of Christ. Official homage to the Egyptian Moses can also be seen in the paintings by Pinturicchio that adorn the Borgia apartments of the Vatican Palace.

But if at first the hermeticists met with a good deal of warmth (both within the Church, and without), they also encountered opposition. To those who disapproved, it was bad enough that Ficino claimed to call down influences from the stars, but Mirandola's claim to angelic magic positively smacked of the demonic. Responding to rumblings of disapproval, Pope Innocent VIII appointed a commission to look into Mirandola's theses, and in 1487 many of them were condemned. Mirandola himself was imprisoned for a time. Two years later, Pedro García, a Spanish bishop on the commission that had examined Mirandola, published a book in which he condemned all magic as contrary to Catholic faith. García did not deny that occult arts were effective; he simply declared that they could not be practiced without diabolical assistance. Finally, in 1563 the Council of Trent took a hard line against magic, and books on the subject were prohibited.

One compounding problem for the magi was the Church's antipathy to the popular magic practiced by village wise women and sorcerers. Although the magi did their utmost to distinguish their intellectual magic from "crude" black magic, they were not wholly

successful in differentiating themselves from lower-class practitioners. Some educated practitioners actually sought out village wise
women and learned from them; notably, Paracelsus, the famed sixteenth-century alchemist, who spent his life championing the common people. For a while, educated interest in the occult helped to
make magic socially respectable, but by the end of the century the
clerical tide was increasingly turning against occult arts of all varieties. The late sixteenth and early seventeenth centuries was the
time of the great European witch-hunts, when it has been esti-
mated that as many as a million women were executed for this
crime.

The Church's reaction against magic (both common and intellectual) was part and parcel of a wider effort to clamp down on all
forms of religious heterodoxy. In the wake of the Reformation,
Europe was seething with all manner of "heresies" and orthodoxy
was being challenged on many fronts. Apart from the schism between Catholics and Protestants, scores of new sects had formed:
Anabaptists, Baptists, Puritans, Quakers, Seekers, Ranters, and
Familists, to name a few. Many of these sects strayed far from the
orthodox fold—most alarmingly, some allowed prominent roles
for women. The Reformation itself had provided the seed for this
innovation when Luther declared that *all* people were equally entitled to commune directly with God. Likewise, the reformers had
encouraged people to read the Bible for themselves, rather than
rely on interpretations handed down by Catholic clerics. But some
of the new sects took the idea of equality in the eyes of God further
than the reformers had intended, and soon, women all over Europe were exercising a more active role in religious life—prophesy-
ing, preaching, and interpreting the words of the Lord.

To Catholics and Protestants alike, women preachers were
anathema, and a backlash ensued. In England, an act of Parliament
passed in 1543 restricted women's right to read the Bible: Only
aristocratic women were allowed to read the sacred texts in private,
merchant class women could do so only in the company of men,
and lower-class people were banned from private reading altogether. In Germany, authorities tried to prevent women from talking about the Scriptures amongst themselves. Both Catholics and

Protestants saw many of the smaller sects as nests of heresy, and, as David Noble has noted, they increasingly associated heresy with women. Clerical determination to clamp down on heresy, particularly by women, created an atmosphere in which religious fervor was easily, and often, mistaken for witchcraft. Historian Carolyn Merchant has noted that lower-class women who behaved oddly, were outspoken, or who challenged male authority were particularly in danger of being accused of being witches. Although, technically, men could also be witches, records reveal that by far the majority of those tried and executed were women. In some villages, almost the entire female population was wiped out.

Socially and intellectually, village witches and hermetic magicians were poles apart, yet both could be accused of trafficking with supernatural forces. To many clerics, *any* such dealings were heretical, and as the seventeenth century dawned, the magi also found themselves increasingly on the defensive. Like the village wise women, they too were increasingly accused of suspect motives. From the Church's perspective, however, the threat of hermeticism was not illusory, because along with practicing magic, a number of prominent magi also preached a message of radical religious reform. Such men wanted to use hermeticism as the basis for a new, and radically unorthodox, form of Christianity. Here indeed was a major heresy in the making.

The religious threat of hermeticism was personified in Giordano Bruno (1548–1600), one of the greatest of all hermetic magi, who was burned at the stake in Italy. Bruno's fanatical hermeticism led him to dream not simply of reforming Christianity, but of replacing it with the ancient Egyptian religion of Hermes Trismegistus. Going further than any of his predecessors, Bruno saw himself as the precursor to a new breed of magician-priests who would usher in a new age of "true" religion. As part of his program of religious reform, Bruno was one of the earliest champions of Copernicus. Yet he supported heliocentrism not for any reason we would now recognize as "scientific," but rather because he saw Copernicus' cosmological diagram as a magic symbol of the world. For Bruno, heliocentrism was a sign that the new age of Egyptian religion was nigh—an age he would personally initiate.

What is astounding about Bruno is that he believed he could launch his radical, and ultimately anti-Christian, reform program from within the bosom of Catholicism. In 1592 he traveled to Italy with the intention of convincing the pope to take up his ideas. This hubris cost him his life. Instead of gaining an audience with His Holiness, Bruno wound up in an inquisitorial prison, where he spent the last eight years of his life before being led to the stake. The irony is that today Giordano Bruno is often portrayed by scientists as a martyr—a man who paid with his life for supporting heliocentric cosmology. However, as historian Frances Yates has shown, it was not his views about science that were the problem. The "genuine" physicists of his own time were as much opposed to his ideas as the clerics themselves.

Although there were practitioners of occult arts who continued to flourish well into the seventeenth century, even within the bosom of the Catholic Church, the climate for such practice was getting ever more dicey. After the death of Bruno, the hermeticist had to walk a fine line to avoid dangerous accusations. It was against this background that the new mathematical scientists sought to define themselves, and one of their overriding concerns was to associate themselves with *orthodoxy*. The self-appointed leader among mathematical men in the fight against magic was the Minim monk Marin Mersenne (1588–1648). Friend of Descartes, champion of Galileo, and a first-class physicist in his own right, Mersenne despised heresy in any form, and from the austere base of his Minim cloister he waged war against the magical worldview.

The Minims were one of the most ascetic orders in France: Their rule required strict obedience to the Church, chastity, humility, and perpetual Lenten fasting. The first three rules were common to all monastic orders, but the last was unique to the Minims, requiring total abstinence from meat, eggs, cheese, milk, and butter. According to David Noble, "members vowed to commit themselves to a rigorous regimen of 'monastic perfection'," and adhered to a strict schedule of religious observances, enforced by correctors who "castigated all lapses." Discipline, obedience, and control characterized the Minim way of life. In short, the Minims were a bastion of Roman Catholic orthodoxy.

For Mersenne, the root of heresy lay in the organic philosophy of nature on which magic was based. In particular, he objected to the idea that the world had a soul that was the source of its vital activity. Originally deriving from Plato, the belief in a world soul all too often led its proponents to unorthodox views about God's relationship to nature. Above all, it led to the heresy of deism. Deists believed that God created the world but that after Creation it functioned purely by natural processes; thus, they rejected the notion of miracles. Others who believed in a world soul equated it directly with God, while for some, the notion of a world soul led to the view that the earth was a living animal. Bruno, for instance, asserted that all the heavenly bodies were huge animals, each with its own individual soul. Mersenne condemned all these notions and regarded Bruno as one of the most evil men who ever lived. That Bruno's hermetic-organic philosophy was coupled with a program for radical religious reform only highlighted for Mersenne how dangerous such thinking was.

During the 1620s Mersenne wrote long, penetrating attacks on every aspect of Renaissance magic and occult art, in which he condemned all the chief perpetrators, including Ficino, Mirandola, and Bruno—by this time all in their graves. But a dead magician was not nearly such a threat as a live one, and Mersenne's most vitriolic attacks were aimed at his contemporary, Robert Fludd (1574–1637). One of the last great magi, Fludd proved a feisty opponent who countered with equally abusive tracts against his attacker. The Mersenne-Fludd controversy was watched with interest by scholars all over Europe.

For his part, Mersenne discounted astral magic, talismans, the powers of plants, stones, and images, and indeed the entire apparatus of natural magic. In its place, he championed a mathematical approach to understanding nature. In the battle against Fludd, Mersenne called on the help of several of his Minim brothers, and finally on the priest Pierre Gassendi, who belonged to an anti-Aristotelian but religiously orthodox intellectual circle in Paris.

Gassendi responded to Mersenne's appeal with an incisive treatise in which he systematically criticized the basis of Fludd's organic world picture; in particular, he challenged the idea of occult forces, and the notion of a vivifying world soul.

Mersenne and Gassendi were not just determined to demolish the organic view of nature so central to the practice of magic; they were equally determined to counter it with a new natural philosophy compatible with orthodox theology. As historian Peter Dear has noted, "The religious significance of [Mersenne's] work was never submerged, always remaining a central motivation." Above all, Mersenne sought to construct a natural philosophy compatible with "orthodox Catholic positions on the relationship of God to His Creation." Here, the Minim monk found his solution in a *mechanistic* conception of nature. Instead of seeing the world as a living organism vivified from within, Mersenne envisioned the world as an inert machine operating strictly according to laws externally imposed by God. "God is not the soul of the world," agreed Gassendi, "but its governor or director."

In Mersenne's eyes, one vital function of a mechanistic philosophy of nature was that it preserved the possibility of miracles. In a strictly lawful universe, not everything is possible by natural process alone, because nature can do only what the laws allow. Anything outside the laws must therefore be achieved by a deliberate act of God. Mersenne understood that the very concept of miracle implied that there must be a natural *order* that could be *broken*. Through mechanism, Mersenne thus meant to counter the heresy of deism. From our perspective in the prosaic twentieth century, it is rather amusing to learn that one of the impetuses for the birth of mechanism was a desire to accommodate the supernatural.

Gassendi's contribution to the mechanistic world picture was to give it a basis in atoms. The world, he said, is not made up of animate objects imbued with souls and alive to occult influences; rather, it is made up of microscopic, inert pieces of matter. This idea was not new, of course: Gassendi had revived it from the ancient Greek philosopher Epicurus. In the Greek form, however, atomism had dangerous atheistic implications, and so, says historian William Ashworth, the pious Catholic priest "tailored an atomic philosophy according to Christian guidelines." In Gassendi's mechanistic world picture, the universe became simply a conglomeration of inert atoms obeying a set of mathematical laws imposed from above by God. The animating spirits and souls of the magical worldview were drained away like the blood from a slaughtered calf. The self-activating universe of the magi was killed off, and in its place stood an inert machine.

Carolyn Merchant has drawn attention to another aspect of the mechanists' agenda. In the Neoplatonic tradition, the world soul was generally regarded as female. By killing off this female power, and vesting all power in a male God, the mechanists effectively negated any notion of a cosmic feminine force. As Merchant has noted, this suppression of the "feminine" in nature mirrored the social suppression of women then being enacted through the witch-hunts. In the early seventeenth century, female power, both in nature and in society, was widely regarded by state authorities, by clerics, and by the emerging scientists as something to be defused.

By constructing a natural philosophy in which the universe was wholly controlled by a male deity, the mechanists created a world picture that reflected the kind of *society* they wished to justify—patriarchal and monarchist. As René Descartes wrote to Mersenne, "God sets up mathematical laws in nature as a king sets up laws in his kingdom." By reconceiving God as the law-making king of the universe, Mersenne and his cohorts consciously defended the royalist state as well as the Roman Catholic Church. "In France," says Merchant, "the rise of the mechanical world view was coincident with a general tendency toward central government controls and the concentration of power in the hands of royal ministries." The political implications of mechanistic philosophy were clearly expressed by another member of Mersenne's circle, the English royalist Thomas Hobbes, who, in his influential book *Leviathan,* proposed "a mechanical model of society as a solution to social disorder." Constructed in opposition to magic, organicism, civil unrest, and rising female power, the mechanistic world picture must be viewed not simply as a product of "science," but also as the by-product of a conservative sociopolitical backlash.

It is one thing to formulate a new philosophy of nature, it is quite another to have that philosophy widely accepted. The mechanists understood that it wasn't sufficient to demolish the foundations of the enemy's edifice; they also had to demonstrate the validity and efficacy of their own approach. They had to show that mathematics was in fact a solid foundation on which to build a new understanding of nature. Living as we do today, surrounded by the fruits of mathematically based science, it is difficult to conceive how little evidence actually existed for this way of seeing in the early seventeenth century. While we now take physics for granted, Mersenne and his fellow mechanists had a far from easy task demonstrating its effectiveness. Mersenne himself launched an early salvo when he compiled an exhaustive list of the things that were known at the time through mathematical science: laws of optics, mechanics, statics, musical harmony, and the like. But of all those who argued the case for a mathematical approach to the study of nature, the most thorough was René Descartes (1596–1650).

Unlike Mersenne and Gassendi, Descartes did not get involved directly in fighting magicians and organic philosophies of nature; rather he put his energies into demonstrating logically that mathematics *was* a foundation for true knowledge. Yet for all his much-vaunted rationalism, at the heart of Descartes' famous argument we find none other than God. Although neither a priest nor monk, Descartes was a devout Catholic, and, along with Mersenne and Gassendi, he too envisaged a science that would harmonize with orthodox Roman Catholic theology. Indeed, he wanted nothing more than to see his own philosophy of nature taught throughout the world by the educational empire of the Jesuits.

Descartes' mission in life became clear to him on November 10, 1619, at the age of twenty-three. He had been traveling in the German countryside and stopped at an inn near Ulm for the evening. There, while warming himself in a stuffy, overheated room, he fell into a reverie that culminated in a vision, in which he was visited by the Angel of Truth. According to historian Edwin Burtt, "the experience can be compared only to the ecstatic illumination of the mystic." The vision was followed later that night by three strange dreams that confirmed through supernatural insight Descartes' already deepening conviction "that mathematics was the sole key needed to unlock the secrets of nature." But because every person cannot be guaranteed a personal visit by the Angel of Truth, Descartes set about to demonstrate the essence of this revelation through logic, finally presenting his argument in his classic text *Discourse on Method*—to which belongs the immortal line: "I think, therefore I am."

In spite of the logical underpinnings of Descartes' text, ultimately the reader learns that it is God's immutability that ensures the existence of mathematical laws of nature. In the Cartesian world picture, it is also the deity's continuing presence that keeps these laws operating from one moment to the next. While Descartes was not an atomist like Gassendi, his matter was likewise devoid of animating spirits and occult forces. It too behaved strictly according to God's laws. Matter, motion, and mathematical law—this was Descartes' parsimonious foundation for reality. There could be no magic in such a stark framework.

Descartes' worldview contained one further element—the human mind—but in his philosophy, mind was entirely separate from matter. Hence his famous dualism of *res extensia*, the material world of matter in motion, and *res cogitans*, the immaterial world of thoughts, feelings, and emotions. This radical separation between mind and matter was to have immeasurable consequences for Western humanity. To put it bluntly: How does the "self" that feels pain and fear and joy fit into the mathematico-material matrix that modern physicists describe?

Frances Yates has suggested that the expulsion of the mind (or psyche) from the world that physicists study can be seen as a further reaction against magic. A basic difference between the magicians and the new mathematical men was that whereas the magician sought to *internalize* the world, to draw it into himself, the mechanist sought to *externalize* the world, to separate it completely from his own psyche. Yates has proposed that "when mechanics and mathematics took over from animism and magic, it was this internalization, this intimate connection of the [mind] with the world, which had to be avoided at all costs." In opposition to the old magical way of knowing, the new mathematical science was to lead not to an emotional, subjective engagement with the world but to a detached and supposedly objective understanding.

Yet if Descartes removed the mind from the world the physicist actually studies, it was he who declared that the human mind *could* triumph over that world. It was the mathematical mind that could discover the true laws of nature, the mathematical mind that would lead to true knowledge of reality. The Dale Carnegie of science, Descartes gave modern mathematical men the confidence that they could truly unlock the secrets of the world.

By the middle of the seventeenth century, magic had been defeated. It is important to reiterate, however, that mathematical men had by no means proven their own approach in any general sense. Their major triumph was still Kepler's planetary laws, and these were still not widely known or understood. Not until the latter half of the seventeenth century would Newton's achievements provide really convincing evidence for the efficacy of mathematical science. The triumph of mechanism over organicism, of mathematical science over magic, cannot be understood simply as a matter of "good" science winning out over "bad," for, in the absence of any definitive evidence, central to the whole battle were issues of heresy and orthodoxy.

The mechanists did not of course win the contest against magic on their own. The chief weapons in this struggle were the powerful heresy-busting machinery of the Roman Catholic Church and the less established but nonetheless effective powers of the Protestants. Yet by providing the clerics with a vision of nature that meshed with orthodox theology, the mechanists rendered the priests an invaluable service. The churches' victory was their victory too.

As secular supporters of orthodoxy, the new physicists were also keen to keep their enterprise male-only. Just as the interpretation of Scripture had long been a pursuit for men only, the mathematical interpretation of nature—God's "other book"—was also to remain a male prerogative. One crucial means by which women were disenabled from participating in the new physics was exclusion from the scientific societies that came bursting into being in the second half of the seventeenth century.

As we have seen, women were already barred from universities, but, as with astronomy, much of the new physics was now evolving beyond academe, in informal networks that gathered around scholars like Mersenne. From the middle of the century, these networks gave rise to official scientific societies, which, almost universally, did not accept women members—a situation that in many instances did not change until the twentieth century. The French Academy of Sciences, which emerged from Mersenne's circle in Paris, was founded in 1666, yet it was not until 1979, more than three centuries later, that a woman was accepted as a full member. That was not, however, for lack of candidates. For the moment it is sufficient to note that in 1911 this august body refused to accept Marie Curie, despite the fact that she had already won a Nobel Prize in physics.

In the seventeenth century, there were not, of course, huge numbers of women trying to participate in physics, because society did not educate many women to the level where they could even begin to think about such subjects. Nonetheless, there *were* well-educated upper-class women who did turn their minds in this direction. Rather than welcoming these pioneers into their new societies, most male members consciously opted to keep the official locus of science female-free. Women could be spectators to the new enterprise, and even (as we shall see in chapter 6) be enlisted as proselytizers, but they were to remain outside the *formal* arena in which science was now discussed, formulated, and honored. Once again, this policy was particularly devastating for women physicists, because, of all the sciences, physics is the one that most evolved from a formalized basis, and within a formalized community.

Furthermore, even though science was now increasingly conducted within a secular environment, there continued to be a quasi-religious undercurrent within the new societies, and particularly among physicists themselves. David Noble and others have argued that the exclusion of women from the early scientific societies can be understood as an aspect of the new scientists' desire to model their institutions along monastic lines. Here we can cite the example of the earliest of these societies, the Accademia dei Lincei (Academy of the Lynx-Eyed), formed under the auspices of the

Italian Prince Frederico Cesi in 1603. Cesi came from a wealthy family with strong ties to the Roman Catholic Church—his brother was a bishop, and among his uncles were a cardinal, an abbot, and another bishop. The Accademia dei Lincei became, in a sense, Cesi's own private religious order. According to historian Martha Ornstein, the original Lincei plan called for the establishment of "scientific, non-monastic monasteries" around the world, and although this missionary expansion never took place, the Lynceans took the monastic ideal seriously.

Their adherence to this ideal went so far as to encompass a commitment to chastity. In the original phase, the rules declared that if a member broke this rule and indulged in intercourse, he had to stay away from other members for three days, and then, with due contrition, beg his brothers to take him back into the fold. Later rules didn't press the point, but the Lincei's official manifesto was filled with exhortations against "libidinous acts," "the attractions of Venus," and even ordinary contact with women. In keeping with this spirit, the Lynceans viewed marriage as a trap that deterred scientific activity. According to Cesi and his fellows—who from 1611 included no less a figure than Galileo— only a "pure" male mind could discover true knowledge. The message was clear: Women were a threat to the genuine seeker of knowledge and were to be avoided as far as possible.

Prior to the formation of the Royal Society in England (perhaps the most prestigious of all scientific societies), John Evelyn, one of the founding members, had proposed to establish a sort of scientific monastery, where "some gentlemen, whose geniuses are greatly suitable," would devote themselves to experimental science. In Evelyn's imagined society, members were to dress in habits and occupy individual cells like monks. There would be prayers, fasts, and communion. As with Cesi's dreams of a worldwide network of scientific monasteries, Evelyn's scheme never came to fruition, but he went on to became a seminal member of the Royal Society, and he gave the organization its name.

The Royal Society did not demand vows of chastity from its members, but a number of its prominent early figures imposed one on themselves. Robert Boyle (1627–1691), a friend of Evelyn's, and the man generally hailed as the "father" of modern chemistry, was an aristocrat who at the age of twenty-one experienced a deep spiritual crisis. Following this critical event he took a vow of chastity for the "love of God," which lasted the rest of his life. David Noble has described him as "an earl in monk's clothing" for whom "[natural] philosophy was a form of worship requiring purification from earthly desires." Boyle's assistant Robert Hooke, who later became secretary of the Royal Society and was one of the greatest physicists of the seventeenth century, also took a vow never to marry. Newton himself, who reigned as president of the society for over twenty years, died a virgin.

Like the French mechanists earlier in the century, the new leaders of English science were anxious to distance themselves from anything that might be construed as heretical—including women. Walter Charleton, another founding member of the Royal Society, summed up many of his colleagues' antipathy toward women when he wrote, "you are the true Hienas, that allure us with the fairness of your skins. . . . You are the traitors to wisdom: the impediment to Industry . . . the clogs to virtue, and the goads that drive us all to Vice, Impiety and ruine." Henry Oldenburg, the society's first secretary, declared that its express purpose was "to raise a Masculine Philosophy . . . whereby the Mind of Man may be ennobled with the knowledge of Solid Truths." This bastion of British science did not admit a woman as a full member until 1945, and, as historian Londa Schiebinger has wryly noted: "For nearly three hundred years, the only permanent female presence at the Royal Society was a skeleton preserved in the society's anatomical collection."

Yet if women were denied entry to the formal locus of the new science, they were not entirely absent from the field. Just as men formed informal scientific networks, so too some women were able to gain access to science through informal channels. Such opportunities were available only to noblewomen, who effectively exchanged access to social rank for access to scientific knowledge. Through her social position, for example, Princess Elizabeth of Bohemia became a correspondent of Descartes. The French philosopher took seriously Elizabeth's objections to his radical dualism, and her questions and criticisms led him to elaborate his views in his *Principles of Philosophy*, which he dedicated to her. Praising her grasp of geometry and metaphysics, Descartes wrote: "The incomparable excellence of your intellect is evident in the fact that in a very short time you have mastered the secrets of the sciences, and obtained a perfect knowledge of them all." Later Descartes was employed by Queen Christina of Sweden, who hired him to draw up the regulations for a Swedish scientific academy and to teach her the new philosophy. Throughout the seventeenth century, there were in fact numbers of women who expressed serious interest in the new physics.

The idea that women might have the potential to make important contributions to science was lent credence by Descartes' own philosophy. His insistence that mind and matter were utterly distinct challenged the Aristotelian view that women were mentally inferior because their bodies were less perfect than men's. If mind operates independently of matter, then nothing about the female body could imply a deficiency of the female intellect. Thus, in the late seventeenth and early eighteenth centuries, Descartes was hailed as a feminist. Yet while he lent his support to individual women, Descartes never used his philosophy for a general defense of the female sex. Though he did regard the sciences as potentially open to women, in his writings he was almost completely silent on the "woman question." Londa Schiebinger has pointed out that this was the case with a number of the more liberal philosophers of the scientific revolution. Both Gottfried Leibniz and John Locke evidenced rather enlightened attitudes toward women, yet, while they lived at a time when the role of women was much debated, neither chose to give it serious philosophical attention.

One man who did, however, was the ex-Jesuit priest and early champion of higher education for women, François Poullain de la Barre. After a standard education in which he was taught the Aristotelian view that women were "monsters," Poullain turned away from scholastic philosophy and embraced instead Cartesianism. He then ventured where Descartes had never dared and applied the French philosopher's method of systematic doubt to the social domain. "Using the tenets of Cartesianism," says Schiebinger,

"Poullain set out to show that there was no significant difference between the sexes. . . . Central to his claim was that the mind—distinct from the body—has no sex." Poullain argued that women were capable of creative work in mathematics, logic, physics, engineering, metaphysics, astronomy, medicine, and anatomy. In short, there was "nothing too high for women."

Unfortunately, such men remained rare exceptions, and, in spite of its promise, Cartesianism did not prove a major force for opening up opportunities for women. While rank could gain a woman informal access to science, nothing could gain her access to the formal inner circles of the emerging scientific establishment—as is vividly demonstrated by the case of Margaret Cavendish, duchess of Newcastle (1623–1673). Born Margaret Lucas, the daughter of a member of the lesser gentry of Colchester, she received little education beyond the type normally deemed suitable for a lady: singing, dancing, reading, and so on. But young Miss Lucas hungered for more, and, recognizing that a woman could gain access to learning only through a man, she chose her husband carefully. In William Cavendish, duke of Newcastle, she found a man who, though thirty years older than herself, was well connected in scientific circles. Soon after their marriage, the Cavendishes were exiled to France (a common casualty for royalists during the English Civil War), and there William drew around him a circle of the new thinkers, which included at times Mersenne, Gassendi, Descartes, and Hobbes. As William's hostess, Margaret gained access to these men around the dinner table.

Yet Cavendish was not interested in being merely a spectator to the scientific revolution. In spite of having no formal education, she wrote six books on natural philosophy, all of which were published privately at her husband's expense. As a thoroughgoing materialist, Cavendish was well-disposed to many aspects of the new mechanism, but she rejected Descartes' radical dualism (as did many people at the time), and she criticized Hobbes' view of matter. Elsewhere she delivered a sharp critique of the contemporary experimentalists, which, although it was not explicitly stated, was probably aimed at Boyle and Hooke. Here then was a woman who sought to engage seriously in the contemporary debates about nature and knowledge, but for the most part her efforts were totally ignored. None of her books was reviewed by the major European journals, and one of the only people who corresponded with her about a scientific matter was the great Danish physicist Christian Huygens, to whom she had sent a set of her works.

The difficulty of Cavendish's position is seen most clearly by the reaction she received when in 1667 she requested permission from the Royal Society to visit one of their working sessions. Given that this wealthy duchess had been a generous patron of Cambridge University, and would have been a major financial asset to the impoverished society, one might have thought the fellows would welcome such a visit. Instead her request caused a furor. After much debate she was given permission to attend a session in which Boyle performed one of his famous experiments. It was to be her only visit. To put this into perspective, it must be realized that in these early days, scientific societies generally accepted noble*men* as members purely for the prestige they would confer. Men above the rank of baron could automatically become members of the Royal Society without the usual scholarly scrutiny. But neither money, title, nor publications could gain a woman entry into this official arena of science.

Women such as Margaret Cavendish, Elizabeth of Bohemia, and Queen Christina are evidence that in spite of their lack of access to formal education, women *were* interested in being part of the scientific revolution. Yet, instead of supporting that interest, for the most part the new mathematical men (and indeed scientists in general) chose to follow the same misogynist path as the medieval academics. As far as women were concerned, the emergence of the scientific societies in the seventeenth and eighteenth centuries paralleled the emergence of the medieval universities in the thirteenth and fourteenth centuries. Nature, like Scripture, remained a male preserve—the domain of Mathematical *Man* alone.

5

$\Omega \quad \mu \quad \delta \quad \Omega \quad \omega \quad \pi \quad \Sigma \quad \Delta \quad \gamma \quad \beta \quad \alpha \quad \hbar$

The Ascent of
Mathematical Man

As a devout Catholic, Descartes had aimed to create a philosophy of nature compatible with orthodox ideas about God and his relationship to the world. Yet, theologians gradually realized that the French philosopher's theory of matter undermined one of the central pillars of Catholic faith—the doctrine of the Eucharist. According to Catholicism, when the faithful receive Holy Communion, the wafer of bread is transformed into the body of Christ. This is the miracle of transubstantiation. But such a transformation was incompatible with Descartes' conception of matter. If Descartes was right, the communicant was not partaking of the body of the Savior but merely chewing on a dry crust; the central act of the Catholic Mass was not a communion with the divine, but the consumption of a flavorless snack. Needless to say, theologians were not enthusiastic about such a prospect, and the Cartesian theory of matter was severely criticized by a number of Jesuits—the very people Descartes had hoped would take his philosophy to the world. In the face of such criticism, he tried to show how his theory harmonized with the Catholic miracle, but after his death the issue was brought to the attention of a papal penitentiary, and in 1663 his *Meditations* was placed on the Index of prohibited books.

Descartes had believed that science and religion could be separated into distinct compartments, wherein each could be pursued in isolation from the other. But in this belief he had failed to recognize the reality of Catholic faith. As the case of transubstantiation demonstrated, by virtue of the fact that both science and theology made claims about nature, they were *intersecting* sets. How then was the relationship between the two disciplines to be negotiated? In the fight against magic, theologians had brokered an arrangement of epistemological power sharing with the mechanists. The clerics had, in effect, agreed to share the interpretation of nature with the practitioners of mathematical science. By accepting the mechanists' world picture as an alternative to the hated organicism of the magi, theologians had tacitly endorsed their science; yet this very science now posed threats to orthodox faith. How then were the new scientists and the Church to deal with potential conflicts about the nature of nature? How much epistemological power were mathematical men going to demand? And how much should theologians concede to them? These issues lie at the heart of the most notorious of all conflicts between Christianity and science—the trial of Galileo Galilei.

Galileo's trial actually took place in 1633, well before the conflict over Descartes' theory of matter, but the issue was essentially the same: To what degree was the Church going to allow mathematical scientists to determine the nature of the world? This was by no means a simple matter of faith against reason, for at the time the Roman Catholic Church was a primary patron of the new mathematical science. In the early seventeenth century, the Jesuit order was at the forefront of research in mathematics, astronomy, and physics. It was a Jesuit, Giambattista Riccioli—not Galileo—who first determined the rate of acceleration of a free-falling body. Despite the popular mythology, the showdown between Galileo and the Inquisition cannot be construed as a simple battle between religious tyranny and scientific rationality. However appealing it may be to think of Galileo as a caped crusader of the intellect, boldly defending truth, justice, and the scientific way, a fuller acquaintance with the story reveals a far less luminous picture. In essence, the Galileo saga was a turf war over epistemological power, in particular, the power to interpret the heavens.

Galileo Galilei was born in 1564 in Pisa, the city of the famed leaning tower, from which legend has it in later life he dropped objects in order to determine that all things fall to the ground at the same rate. Although there is no evidence he did anything of the sort, like so much else about Galileo's life, the story has become part of modern scientific mythology. Hints of Galileo's destiny were already evident in the life of his father, Vincenzo, an impoverished member of the lower nobility of Pisa. A learned man with radical leanings, Vincenzo was an accomplished composer and a student of music theory who made a significant contribution to the mathematical study of harmony. Vincenzo gave his son an excellent education, but because of the family's poor financial circumstances, he dreamed of the lad becoming a merchant. As the first son, Galileo would be responsible for providing dowries for his three sisters. Yet, to his credit, Vincenzo recognized his boy's unusual ability, and instead of condemning him to a life of commerce, sent him to the University of Pisa, ostensibly to study medicine. By this time, however, young Galileo had fallen in love with mathematics.

Despite his obvious ability, Galileo was not immediately destined for a brilliant academic career. The university refused to grant him one of the scholarships it retained for poor students and he eventually left without receiving a degree. That Galileo did not get a scholarship is worthy of note, because while at university he had already made an important scientific discovery—that a pendulum of a fixed length always swings with a constant frequency. Given this, some historians have viewed the university's omission as an early bid by the old guard to thwart a revolutionary thinker. But there may have been nothing so sinister afoot, for, along with his scientific genius, Galileo had a genius for making enemies. He may simply have alienated the scholarship selectors with what one biographer has described as his "cold, sarcastic" personality. In any event, the young physicist was undeterred by this lack of recognition from academe and continued his work at home, where he soon became absorbed in studying and designing mechanical devices. As would also be the case with Newton, Galileo had an innate flair for gadgetry, and his youthful inventions included the "thermoscope," a forerunner to the thermometer, and the "pulsilogium," a device for timing pulses.

Galileo circulated reports of his mechanical researches in manuscript form, and scholars soon began to take notice. In the great Renaissance tradition, he also began to seek patrons. One of the earliest and most loyal was Cardinal Del Monte. Through Del Monte's help, at the age of twenty-five Galileo was granted the post of lecturer in mathematics at the University of Pisa, and three years later, again with the help of the cardinal, he was appointed to a professorship of mathematics at the University of Padua. It was during this Paduan period that he did much of the work that would lead Einstein to call him the father of modern dynamics.

Although Galileo was laying the foundations for a new physics of motion, he only shared his ideas intermittently with a few select correspondents. It is one of the greater ironies of the history of science that while Galileo has come to be seen as a champion of radical new ideas, for most of his life he kept his own to himself. He did so not for fear of religious persecution but because he did not want to risk the ridicule of his fellow professors, most of whom were still committed Aristotelians. For all his reputation for boldness, Galileo was afraid of being laughed at. Despite the fact that he was a convert to Copernicus, in public he continued to lecture on Aristotelian physics and Ptolemaic cosmology.

In 1609, however, a new gadget would propel Galileo out of the cosmological closet. In that year, he learned about an instrument Dutch opticians had invented consisting of two lenses mounted inside a hollow tube, through which objects could be seen at a distance. Quickly grasping the principle behind it, Galileo constructed his own from lenses acquired at a local spectacle maker's. But this crude instrument could magnify objects only three times, so he set about improving the design himself. To that end he had to learn to grind his own lenses, a task requiring much skill and patience; but by doing so he made an instrument that could magnify fully nine times.

Always one with his eye on the prize, Galileo demonstrated this marvel to the city fathers of Venice. With it, one could see ships approaching two hours before they could be seen with the naked eye. Would not such an instrument be an invaluable aid to the defense of their waterbound city? Allowing a few days for the point to sink in, Galileo then presented his "optick tube" to the city as a gift, along with a letter hinting that his scientific researches were much in need of funding. The continuation of his work, he subtly suggested, would lead to even greater marvels from which Venice might benefit. Anxious not to miss out on whatever technological revolution might be afoot, the city fathers commuted his position at the university to a lifetime appointment and doubled his salary. Much to the amusement of Venetian society, similarly powerful instruments were soon available for a few scudi in local shops.

Having attended to business, Galileo did something seemingly very innocent: he turned his spyglass to the stars. Therein he begat a revolution. Galileo was not the first person to use the Dutch spyglass as an astronomical instrument. A few weeks before, Thomas Harriot, an English scientist, had examined the moon through one. But if Harriot had the temporal advantage, it was Galileo who first understood the true significance of the soon-to-be-named telescope. Overnight the field of cosmology had changed—no longer just a matter of people using their minds to make sense of what their eyes told them; now humanity would also have to deal with the mysteries that came flooding down the "optick tube."

What Galileo saw through his telescope sent shock waves through Europe. Most momentously, he found four moons orbiting the planet Jupiter. Today astronomical discoveries have become so commonplace it is difficult for us to imagine what a singular event this was. One must remember that at the beginning of the seventeenth century, astronomers knew of no more celestial bodies than the ancient Sumerians. Until the advent of the telescope, the sky contained one sun, one moon, five planets, and a fixed array of stars. Suddenly to discover four new members of the celestial family was almost as momentous as to discover four new continents, and, not unjustly, Galileo saw himself as the Columbus of the stars.

Furthermore, Galileo's discovery provided powerful circumstantial evidence in favor of Copernicus, because the Jovian moons proved that not *all* heavenly bodies revolve around the earth. They did not prove that anything orbited the sun, but they did show that the earth was not the center of everything, as Aristotelians maintained. But if the earth was not the center of *everything*, then why should it be the center of *anything*? Psychologically, the moons of Jupiter were a major blow to champions of geocentrism, who still constituted the majority. Even more devastating, in the long run, was Galileo's further discovery that Venus has phases like the moon. Here at last was *direct* evidence that at least one planet *was* orbiting the sun.

Galileo informed the world about his discoveries in a deft little book entitled *The Starry Messenger*, published just a year after Kepler's *New Astronomy* announced *his* discovery of ellipses. But whereas most people, including Galileo, were unable to grasp the significance of Kepler's achievement, Jupiter's moons made Galileo an instant celebrity. Anyone who could get his hands on a decent optick tube could see for himself the Jovian satellites, and the waxing and waning of the planet of love. (Actually, because the telescopes of the day were of such poor quality, it was rather difficult to see anything at all through them. One historian has quipped that the miracle was not that Galileo found the moons but that he could even find Jupiter.) With his new-found celebrity, the Great Discoverer of the Moons of Jupiter was soon appointed Chief Mathematician and Philosopher to the court of Cosimo di Medici (the Great), whom Galileo had cannily honored by naming the moons the Medician Stars. After his telescopic discoveries, Galileo came to see himself as the voice of the new astronomy. At one point he even claimed that *all* discoveries made with the telescope belonged to him alone. Most significantly, after decades in the cosmological closet, Galileo "came out"; from now on he would be increasingly vocal in support of Copernicus.

Nonetheless, fully two decades would pass before the matter came to a head in Galileo's infamous trial. The catalyst for this event was not science itself but politics and personality. As historian Mario Biagioli has shown, Galileo's career must be understood within the context of patronage dynamics. As a "client" of the Medici, he was expected to provide amusement for Cosimo and his courtiers in the form of scholarly disputes. In Baroque courts, such disputes were seen as the intellectual equivalent of knightly jousts, and Galileo was Cosimo's knight of the mind. As with medieval knights, it behooved him to put his valor and skill at the service of his lord on the intellectual field of honor. With a tongue as sharp as his mind, Galileo proved adept in this role, and could eviscerate rivals in the most subtle, yet amusing ways. One particularly juicy joust involved the question of why certain bodies float on water, while others sink. Against his Aristotelian opponents, who insisted that bodies float or sink depending on their shape, Galileo championed Archimedes' view that it depended on their density.

Unfortunately, many of his rivals in these intellectual jousts were leading Jesuit scientists, and Galileo spared them no mercy. Indeed, he relished the opportunity to make fun of clerics, who, in the minds of the courtiers, were the epitome of stuffy pedantry. He was particularly vitriolic against Jesuits who opposed Copernicus. But the desire to entertain the ladies and gentlemen of the Medici court, combined with a monumental ego, led Galileo to tactics that were sometimes far from amusing. In one infamous episode he falsely claimed priority for the discovery of sunspots, and then accused a Jesuit scholar who had seen them beforehand of stealing his ideas. With such tactics he eventually succeeded in alienating the entire Jesuit order, an extremely powerful body of men who had once been his supporters. Given that Galileo hoped to gain the support of the pope, this was a particularly foolish strategy—the Jesuits were, after all, the pope's "soldiers."

In 1623, when his old friend Cardinal Maffeo Barberini ascended to the papal throne as Urban VIII, Galileo had the perfect opportunity to plead the case for heliocentric cosmology at the highest level. Urban was a sophisticated man, well versed in the new science, and he assured Galileo that he was free to write openly about heliocentrism. But he, like everyone else, was to do so only *hypothetically*. He was not to imply that it was a proven fact. Over the next decade, Galileo became increasingly bold in flaunting this accepted practice, and more and more he spoke of heliocentrism as if it had been unequivocally proven. Finally, in 1632, he published a book in which he appeared to make fun of the pope for failing to jump wholeheartedly onto the Copernican bandwagon. Urban was deeply offended, and the recalcitrant physicist was soon summoned to the Holy Office for interrogation.

By the time of his trial, Galileo was sixty-nine years old. Throughout his life he had counted churchmen among his most ardent supporters, and, despite his tactlessness toward the Jesuits, he still had powerful ecclesiastical friends. Various clerics tried to work out a compromise that would have prevented a trial, but Galileo stubbornly repelled their efforts. However, because he was a famous philosopher, and not some village witch, he never spent a day in jail and was housed in luxury during his hearings. On the final day of the trial, the Inquisitors knew beforehand that Galileo had decided to retract his forbidden statements. The threat of torture that was read to him was purely a legal formality. This does not of course make the action justifiable, but it is important to understand that the whole business was a lot less dramatic than is often imagined. There was no one busy in the basement stoking a pyre or oiling the rack. On the appointed day, Galileo formally abjured his beliefs, and the inquisitors let him go.

That the Church never intended to hurt Galileo or to interfere with his scientific work is shown by the fact that his punishment was to spend the last years of his life under house arrest at his own villa. During this time he wrote the book that established his real position in the history of physics, the book in which he laid out the foundations of modern dynamics, the mathematical study of motion. While house arrest is certainly a curtailment of liberty, it hardly constitutes a war against science. Those who bristle at the treatment of Galileo should spare a thought for the hundreds of thousands of women burned at the stake as witches. In the hands of modern champions of science, the story of Galileo has been distorted into a self-serving myth that does injury to both the Church and science and glosses over what was really at issue.

Central to the whole affair was the issue not so much of truth, but of proof. How much proof should the new scientists be required to produce before the Church ceded authority over the heavens to them? Copernican cosmology was incompatible not only with biblical passages that claimed the earth as the center of the cosmos, but also with direct human experience. When we look up at the sky, the celestial bodies *do* appear to be revolving around us. Geocentrism was not an artifact of Christian imagination, but a logical deduction from the evidence of the senses. It was heliocentrism that was the imaginative fable. Nonetheless, Copernican cosmology had slowly been gaining supporters since the mid-sixteenth century—quite a few of them members of religious orders. In that time, the Church and Copernicans had come to an arrangement whereby those who wished to talk about a sun-centered system would do so under the banner of "saving the phenomena." They would make a disclaimer that they did not mean to assert the earth *really* moved round the sun, they only wished to employ heliocentrism as a useful and convenient way of describing heavenly motions. In effect, Copernicans acknowledged that heliocentrism was not a proven fact, but only a *hypothesis*—and, given the evidence available at the time, that is precisely what it was.

But Galileo decided that he was not happy with saving the phenomena. He wanted to say that heliocentrism was a *fact*. Moreover, he wanted the church to acknowledge this, and he tacitly suggested that those passages of Scripture that conflicted with Copernicus ought to be interpreted metaphorically rather than literally. Not surprisingly, many theologians were not impressed at being told by a layman how to treat such an important theological matter as scriptural interpretation. Whatever legitimacy science might have did not extend to biblical exegesis.

If Galileo had definitive proof that the earth revolved round the sun, the situation would have been different. But, although the phases of Venus provided evidence that *it* was orbiting the sun, there was no direct evidence of the earth's motion. Indeed, quite the contrary, because, according to the physics of the day, if the earth *was* moving, then people ought to have observed things they hadn't—for instance, arrows ought to have flown differently depending on which direction they were shot. Scientifically speaking, there were very good reasons to think the earth was perfectly still. The inquisitors made it clear that if Galileo could produce definitive evidence, they would reconsider their position, but he couldn't. In the end, he had to admit that heliocentrism was only a hypothesis. This, in sum, was the outcome of the celebrated trial. Galileo wasn't asked to stop thinking about heliocentric cosmology, or its consequences for terrestrial physics; he was simply forced to acknowledge the truth about its status as a theory.

In the early seventeenth century, the Roman Catholic church was not against mathematical science. It was simply under no obligation to cede authority over the heavens to mathematical men in the face of such scanty and ambiguous evidence. As historian Edwin Burtt has remarked, "Contemporary empiricists, had they lived in the sixteenth century, would have been the first to scoff out of court the new philosophy of the universe." The point is that Galileo tried to claim for mathematical science more than it could support at the time. That history ultimately revealed heliocentrism to be valid is no argument for concluding that the Church was wrong. To demand concrete proof of a radical new theory is not an act of tyranny but good *scientific* practice. Scientists themselves demand no less.

Ironically, Galileo's ego led him to reject by far the most convincing evidence for a heliocentric cosmos that existed at the time: Kepler's ellipses. Here was hard empirical evidence for a sun-centered system. But Galileo was not a man to push forward someone else's discovery; besides, he personally rejected the ellipses. Looking back, we can see that it is an enormous pity these two men did not combine forces, as Kepler dearly wished. What a powerful team they would have made! The deeply religious Kepler recognized the danger of Galileo's confrontational tactics and understood that the Church needed time to make such an important shift in its thinking. Like Descartes, Kepler believed that, if given time, theologians would come around of their own free will, and indeed there are good reasons for thinking that might well have been the case had Galileo not forced a showdown.

The effect of Galileo's actions, and the trial they precipitated, was to create in Italy a climate of distrust between science and theology that dampened the development of physics in that country for the rest of the century. But whatever the legacy of Galileo in Italy, Christianity and the new physics were *not* inherently incompatible. In 1642, the year Galileo died, the greatest mathematical man of all time was born—a man who would prove once and for all the soundness of heliocentrism, and who would weave together physics and Christianity in a glorious synthesis. If ever one needed proof that physics and religion are not necessarily at odds, one need look no further than Isaac Newton. Whereas Galileo sought to wrest epistemological power away from the clergy, Newton sought to share it with them. Whereas Galileo saw the clergy as his enemies, Newton saw them as his allies. Following firmly in the tradition of Pythagoras, the man who finally established physics as "the queen of the sciences" saw his entire life's work as a search for God.

In contrast to Galileo, Newton's family provided no preparation or hint of the man he was to become. Until the arrival of young Isaac, clan Newton (at least on his father's side) was illiterate. Isaac Senior couldn't even sign his own name and had formalized his will with his thumb print. But if the Newtons couldn't read, they certainly had a knack for business; for several generations the family had become steadily more prosperous growing grain and raising sheep. At the time of Newton's birth, he was heir to a country manor, and his parents fully expected he would one day succeed his father as the head of their large farm. But fate had other plans for this country boy.

When Newton was only three months old, his father died, and three years later his mother, Hannah, remarried. Her new husband, a parson from a neighboring village named Smith, had no intention of taking the son along with the mother, so, while Hannah moved away, Isaac was shunted off to his grandmother. It is hard to imagine that this loss could have been anything less than traumatic to the child and that it did not play a significant part in the formation of his character, which above all can be summed up by the word *loner*. All his life Newton was a singular, antisocial person who made few friends and many enemies. Of his fellow students at Trinity College in Cambridge, not one recorded ever having any contact with him. His record with women was even more bleak. The only time he is known to have shown any interest at all was an early teenage crush on the daughter of his grammar school headmaster.

Newton's biographer Richard Westfall has suggested that even as a child his precocious mind was isolating him from his peers. Instead of playing roughhouse games with his school chums, young Isaac spent his time making intricate working models of windmills and waterwheels. As with Galileo, Newton's self-styled mechanical apprenticeship helped prepare his mind for a mechanistic philosophy of nature. More telling even than his model mechanisms were the sundials Newton constructed. From an early age he had been fascinated by the sun, and he filled the house with homemade devices for telling the time from shadows. This solar love affair lasted Newton's entire life, and as an adult he knew the shadows in every room of his own home so well he could tell the time simply by glancing at them. As Westfall has pointed out, this boyhood tinkering with sundials not only sparked an interest in the study of light (to which Newton would later make major contributions) but also gave him an intimate awareness of the cosmic order. By charting the daily and yearly passages of shadows, young Isaac experienced in a tangible way the regularity and patterning of the great celestial mechanism. Even as a child, with no formal training in mathematics or science, his mind was reaching out to the universe beyond.

As an adult, Newton was as prolific as he is famous, and a vast collection of his papers has survived. Throughout his life he filled notebooks with his forays into science and mathematics, and historians are still mining this treasure trove, discovering the full, exhausting extent of his achievements. But along with the "legitimate" science, Newton left behind half a million words on alchemy, a subject to which he ultimately devoted more of his time than to physics. Newton was extremely secretive about this side of his activities, which was in keeping with the alchemists' view of themselves as an elite illuminati who must shield their knowledge from unworthy minds. But, as he well understood, in the antimagical climate of the late seventeenth century, such secrecy was also prudent. Ever cautious about how he might be perceived, Newton never publicly talked about alchemy, and toward the end of his life did his utmost to perpetuate a strictly orthodox view of himself. Historians followed suit, and for almost three hundred years chroniclers of science swept Newton's alchemy under the carpet. But, in the last few decades, there has been growing recognition that this "heretical" sideline was a central facet not only of his personal life, but also of his life as a scientist. Mathematician and magician, physicist and alchemist, no one has stepped more firmly on the "true" path and yet made so many excursions into the arcane wilderness as Isaac Newton.

Newton's formal introduction to science occurred during his student days at Cambridge. Yet the university itself offered no instruction in the new physics. Still entrenched in an Aristotelian curriculum, the dons of Cambridge were not in touch with the revolution in natural philosophy taking place on the continent. Once again, Newton was compelled by his own drives, and he found his way to the works of Kepler, Descartes, and Galileo. Genius is always in part a matter of timing, and no genius has ever had better timing than Newton. It is probably fair to say that no scientist before or since has had the advantage of such exceptional immediate predecessors. All three left a legacy bursting with fruitful directions to explore, and Newton made the most of this abundance by building on the works of them all.

If Newton undoubtedly stood on the shoulders of giants, it is undeniable that he was also one himself. Within a year of study, he had mastered the entire corpus of seventeenth-century mathematics and launched into the development of calculus. At the same time, inspired by Galileo and Descartes, he had begun to investigate the physics of motion, and had quickly advanced from eager student to the forefront of the field. By the age of twenty-five, the Lincolnshire farm boy had not only become the foremost mathematician in Europe but also equaled its greatest physicists, and he was soon to surpass them all.

The turning point came in 1666, a year Newton would later call his "anno mirabilis." Because of a heavy bout of plague, Cambridge University had closed, and the budding philosopher of nature was spending the hiatus on his mother's farm. There, amid the hedgerows and sheep, surrounded by rustic relatives to whom he appeared an uncommonly useless young man, Newton continued his private investigations into the foundations of physics. One day while he was musing in the garden, the fall of an apple from a tree caused him to wonder if the power of gravity which pulled the fruit to the ground might not extend beyond the earth. Could it extend to the moon? A quick calculation based on a formula he had recently discovered for circular motion showed him there was indeed a correlation between the force of gravity here on earth and the motion of the moon around our planet.

Contrary to the Newtonian legend, however, the universal law of gravity did not spring fully formed into young Isaac's mind with the fall of the apple. It would be twenty years before such clarity emerged. This lengthy gestation in no way suggests that Newton was less than the legend portrays; it only reveals that understanding gravity required far more than a single flash of insight. As Richard Westfall has commented, "[the great French mathematician] Lagrange did not call him the most fortunate man in the universe simply because he had a bright idea." As is so often the case in science, inspiration would have to be supplemented by a great deal of perspiration.

Aside from the scientific problems, there were also great psychological barriers to be overcome. One major stumbling block was that, in the late seventeenth century, the idea of an invisible force acting across empty space was anathema to any self-respecting scientist. Having fought so hard against the magicians and their heretical occult forces, the new mathematical men were hardly likely to be sympathetic to the idea of an eerie force emanating from celestial bodies. Newton had originally been a committed mechanist, but he gradually came to realize that there *must* be some kind of force holding the planets in orbit around the sun, and the moon in orbit around the earth. This celestial situation is analogous to a stone being whirled around on a string—if the string breaks the stone will fly away. With the string one can see what holds the stone in orbit, but with the celestial bodies, there was no visible evidence of a restraining power. Nonetheless, Newton became convinced that one must exist. Taints of astral magic notwithstanding, the physicist in him realized that nature *did* manifest an eerie celestial force.

Historians now believe that Newton's acceptance of this heretical idea in an avowedly mechanistic age owes much to his immersion in alchemy. It is not insignificant that the only other founding father of modern physics who embraced the notion of an invisible celestial force was Kepler, a devotee of astrology. Galileo, Descartes, and Leibniz all rejected the idea, and on the continent, Cartesians continued to ridicule gravity well into the eighteenth century. But personal experience of occult arts allowed both Kepler and Newton to accept something their supposedly more rational peers could not. Through this "magical" notion, the science of physics was advanced.

Once Newton had accepted the possibility of an invisible force, he assembled the pieces of the gravitational puzzle in a series of brilliant but breathtakingly simple steps. He showed how the tendency of things to fall to the ground on earth could be explained by the same force that holds the moon in orbit around the earth, and how that force could, in turn, explain the orbits of the planets around the sun. Thus, a single power could account for terrestrial, lunar, and planetary action. Gravity, said Newton, is a force that radiates out from every massive body and draws things toward it— be they apples, moons, or planets. Newton did not give the world just the idea of a universal gravitational force; he also gave us an *equation* that describes precisely how this force behaves. This universal law of gravity is one of the simplest, most elegant, and most powerful equations in the history of science. In half a dozen symbols, heaven and earth were united, and heliocentrism at long last received a firm mathematical foundation. After a century and a half, the transformation begun by Copernicus, and championed by Galileo, had been completed.

Yet for all that Galileo and Copernicus are hailed as the heroes of heliocentrism, it was Kepler who provided the crucial groundwork for Newton's synthesis. In the first instance, Kepler's planetary laws gave Newton clues about the nature of gravity; and in the second, once Newton had worked out his gravitational equation, Kepler's laws served to verify that it was valid. All of the planetary laws that Kepler had discovered turned out to be logical consequences of Newton's law of gravity. It was this wondrous agreement between observation and theory that eventually convinced people that, whatever their objections to occult forces, Newton must be right. And, in accepting gravity, people were also compelled to accept a heliocentric system. Thus, Newton's law marked a turning point for Western science: With this equation, authority over the heavens finally passed from theologians to physicists.

Ironically, heliocentrism came to be accepted without any direct evidence that the earth itself moved. Not until the nineteenth century would definitive evidence be available. Thus, the inquisitors' demands of Galileo had still not been met. In this respect, the universal law of gravity marks an even more significant point in the history of science, for, in place of concrete *physical* evidence, people accepted the testimony of an *equation*. From now on, the mathematical relationships that physicists discovered would serve not just as descriptions of phenomena but increasingly as primary sources of insight into nature. At last, the inheritors of Pythagoras were besting those of Aristotle. With Isaac Newton, Mathematical Man ascended the epistemological throne.

If Newton had done nothing else, his place in history would still have been assured. But along with the law of gravity, he also discovered the three laws of motion, which for over two hundred years remained the archetype for scientific laws. The law of gravity is, in fact, just a special case of the second of these laws. Newton's laws of motion apply not just to celestial bodies but to all material bodies. They are to physics what the Ten Commandments are to Christianity: the basic principles that are supposed to govern all action. This economical trinity begins with the law of inertia, which states that a body will continue in the same state of motion until it is acted on by a force. The second is the law of force, which states that when a body is acted on by a force its motion changes in inverse proportion to its mass. This is encapsulated in the simple equation: Force = mass × acceleration. The much quoted third law states that for every action there is an equal and opposite reaction.

Newton presented his laws of motion and gravity in his scientific masterpiece, *Mathematical Principles of Natural Philosophy* (1687), commonly known as the *Principia*. It was the most important book in the history of Western science since Aristotle's *Physics,* and during the eighteenth century his achievements came to exemplify science itself. People in a wide variety of fields sought to emulate him and to become Newtons in their own areas. Philosophers formulated basic laws of thought and talked about "a kind of attraction" between ideas; chemists tried to explain chemical reactions in terms of a gravitylike force of attraction between substances; and physicians tried to show how disease and health could be represented in terms of Newtonian forces.

According to historian Derek Gjertsen, Newton's own doctor, Richard Mead, saw health as a function of the state of the body's fluids and believed that ultimately "treating a patient would require little more than solving a set of hydrodynamic equations." Even the social sciences were invaded by this Newtonian impulse. In *The Spirit of the Laws,* an eighteenth-century sociologist emulated the method of the *Principia* and attempted to derive by logical deduction from a set of fundamental principles consequences of a sociological nature. No modern scientist except Charles Darwin has had such an immense impact, not only on his own science, but on society at large. And, despite the immense advances of twentieth-century physics, the Western psyche still operates within a largely Newtonian framework.

The possessor of this great mathematical mind was also a profoundly religious man. Along with his scientific and alchemical writings, Newton penned a vast body of theological writing. Much of it was devoted to the interpretation of biblical prophesies, a subject with which Newton was obsessed, and, as with everything he turned his mind to, he brought to this activity a relentless vigor and thoroughness. He taught himself Hebrew so that he could read the Scriptures in their original language and thereby be better equipped to ferret out the truth behind the prophets' words. For Newton, God was not only the creator of the universe but an active and continuing Governor of the world who worked through history as well as nature. By correct interpretation of biblical prophesies, he aimed to demonstrate God's mastery over the affairs of humankind. This incomparable physicist also devoted considerable energy to determining the precise dimensions of the Temple of Solomon—a sacred construction of antiquity that supposedly represented a blueprint for heaven. Whereas Galileo had been interested only in the book of nature, Newton studied the Scriptures as thoroughly as he studied the celestial motions.

Newton's respect for the written word of God was also manifest in his respect for the clergy—in his case, Anglican rather than Catholic. Whereas Galileo had delighted in poking fun at clerics, Newton's intellectual development was deeply influenced by a group of Cambridge-based theologians, and later in life, when he was trying to promote his natural philosophy on a wide scale, he looked to the clergy as his primary allies. He in turn became theirs. Following in the footsteps of Kepler, Descartes, and Mersenne, Newton wanted nothing so much as to put his science at the service of theology. In a famous reply to a request from Anglican theologian Richard Bentley, who had written to inquire if it might be possible to use the *Principia* as an argument for God, Newton responded: "Sir; when I wrote my treatise about our system, I had an eye upon such principles as might work with considering men, for the belief of a Deity; and nothing can rejoice me more than to find it useful for that purpose."

Newton wanted his science to be not merely compatible with religion but to reinforce it. To this end he saw his own natural philosophy as an antidote to the rather bleak mechanism of Descartes. Although Descartes had formulated his philosophy of nature with a view to serving the Church, by the late seventeenth century, many people had come to see his brand of mechanism as a recipe for atheism. Among those who did were the Cambridge

Platonists, the group of liberal Anglican divines who influenced the young Newton. As Anglicans rather than Catholics, the Cambridge Platonists were concerned not with the problem of the communion wafer but rather with the fact that Descartes appeared to have written God entirely out of the universe. While the French philosopher had acknowledged that God's continuing presence was necessary to sustain the laws of the universe from one moment to the next, he had insisted that God did not *design* the universe, and that, furthermore, once it had been established, God did not intervene within it in any way. The Cambridge Platonists found it difficult to see how such a remote and abstract conception of the deity could offer a foundation for a *moral* relationship between humanity and its maker.

In opposition to that of Descartes, Newton's own natural philosophy was grounded in the belief that God was both the providential designer of the universe and its active and beneficent overseer. Having expressed the view to Bentley that nothing could rejoice him more than to find that his scientific work could be used to demonstrate the existence of a purposeful and caring deity, Newton went on, in a remarkable series of letters, to elaborate just how this might be done.

First and foremost, in his eyes, was the evidence of the solar system itself. According to Newton, no natural cause could have created a system in which all the planets moved in the same direction, in the same plane, and in concentric orbits. Such systematic order, he believed, could spring only from the providential action of a supernatural power. "This most beautiful system of the sun, planets and comets," he wrote, "could only proceed from the counsel and dominion of an intelligent and powerful Being." Just as the solar system evidenced need for a cosmic designer, to Newton, it also demanded a cosmic maintainer. Because every single body exerts a gravitational pull on every other body, Newton believed that many years of cumulative small tugs on the planets—caused by other planets and by passing comets—would eventually destabilize the whole system. Thus, he believed that God would be required to step in now and again and set things back to their proper place. God could be seen, in effect, as a cosmic watchmaker, constantly caring for and readjusting his celestial mechanism. Such intervention need not violate the laws of nature, however; Newton suggested that God could employ comets for this purpose.

Everywhere Newton looked he saw evidence for a providential and active God—not only in the heavens but also on earth. He once asked: "Whence is it that nature does nothing in vain and whence arises all that order and beauty which we see in the world?" For him the answer was definitively God. "All the diversity of natural things which we find suited to different times and places could arise from nothing but the ideas and will of a Being necessarily existing." But Newton's God did not oversee the maintenance of the world from some remote pinnacle; unlike Descartes' deity, Newton's divine overlord was present throughout the material world. He achieved this omnipresence through the medium of space, which, for Newton, was nothing less than God's sensorium. By his omnipresence (mediated through space), God was all-seeing, all-discerning, and finally, all-ruling. In Newton's words: "He is eternal and infinite, omnipotent and omniscient; that is, his duration reaches from eternity to eternity; his presence from infinity to infinity; he governs all things, and knows all things that are or can be done." In short, as the absolute Governor of Creation, Newton's God was the very antithesis of Descartes'. Rebelling against the remote uncaring deity of Cartesianism, Newton returned God to intimate daily involvement with the material world.

As well as bringing his science to the service of theology, Newton allowed his theology to influence his science. Most important, he argued that space must be *absolute* because it was synonymous with the presence of an absolute God. For him absolute space (and time) became metaphysical axioms. This theologically inspired conception of space (and time) would quickly become one of the mainstays of modern physics, and long after physicists ceased to associate space with God, they retained a Newtonian insistence on an absolute framework for reality. Yet it was precisely this conception of space and time that Einstein would have to repudiate. Newton's ideas about space reveal how intimately his physics was entwined with his religious thinking. As Edwin Burtt has put it,

Newton's religion was no "mere appendage to his science" but "something quite basic to him."

If anything, it was Newton's science that was an appendage to his religion. In the course of his life, the physics emerges as just one part of a wider program whose ultimate aim was entirely religious. Beyond Newton's desire to serve Anglicanism lay the deeper desire to reclaim what he saw as an original and pure form of Christianity. To quote historian Penelope Gouk, "Newton regarded his natural philosophy as an integral part of a radical and comprehensive recovery of the true ancient religion." According to him, this true religion had originally been revealed by God to Noah. Contained in this pristine Christianity had supposedly been a true knowledge of the world, which Newton believed had been passed down from Noah to Moses, and from him to the Egyptians and Greeks, notably Pythagoras and Plato. Newton's ultimate aim was to recover that lost Adamic knowledge, and his private notes reveal that he saw himself not as a pioneer but as a *restorer* of the ancient wisdom God had given to humankind.

Newton assembled a large array of evidence in an attempt to prove that the ancients had known much of what he presented in the *Principia*. One intriguing piece of "evidence" was his assertion that Hermes Trismegistus had been "a believer in the Copernican system." That Newton in all seriousness should quote the authority of Trismegistus almost a century after the Egyptian sage had been revealed as a fiction is a clear indication of the deeply unorthodox nature of his thinking. More substantially, he asserted that Pythagoras had known the universal law of gravity, and he concocted a convoluted argument to demonstrate that this law was in fact the true knowledge concealed in the Pythagorean idea of the harmony of the spheres. Rather than look forward to some innovative future, Isaac Newton set his sights firmly on the distant past.

His desire to recover the lost Adamic knowledge was not for the sake of the knowledge itself but, above all, for moral reform. As historian Piyo Rattansi has explained, Newton believed "once human beings comprehended the infinite power of God, and how He had framed things and continually watched over them, they would be led to a deeper understanding and acceptance of the duties they owed to Him and to their fellow human beings. The recovery of the true scientific account would thus be followed by a restoration of the true morality since it was founded on a genuine conception of God and his providence."

In Newton's belief that true (scientific) knowledge of the world would serve the cause of moral reform, we see distinct shades of Giordano Bruno. He too sought to restore true morality through true (magical) knowledge of the world. But whereas Bruno was a heretic, Newton was a good Anglican. Or was he? Although publicly he lent his science to the Anglican cause, privately he too held views that were nothing short of heretical. In particular, for Newton a return to the true religion meant the rejection of the doctrine of the Trinity—the belief that God, Christ, and the Holy Spirit are one tripartite divinity. Trinitarianism is one of the central dogmas of both Catholic and Anglican faith but Newton believed it was a corruption of the true faith and that God alone was fully divine—the implication being that Christ was not. Theologically speaking, rejection of Trinitarianism (or what is known as Arianism) was a serious crime; so serious that had Newton's real beliefs been known, he would not have been allowed to remain at Cambridge.

Fortunately, Newton never deluded himself that he could be open about his unorthodox views, and he revealed his secret only in his private notes. But if he did not go public with his Arianism, neither was he a hypocrite. He was quite prepared to forgo an academic career rather than lie about his religious persuasions. During Newton's time fellows of Cambridge still had to be ordained ministers of the Anglican Church and therefore had to declare formal adherence to the Church's doctrines, including Trinitarianism. Newton was not prepared to do this, and in 1675 he was expecting to leave the university. But as the deadline for ordination approached, a dispensation was granted by the king: Isaac Newton could be a fellow without being a minister.

Quite apart from revealing the full depth of Newton's religious spirit, this incident raises the not inconsequential question, What if he had not received the dispensation? What if he had not stayed on at Cambridge but had returned instead to the farm in Lincolnshire to live the life of a country squire? Would there have been a *Principia,* one grand, embracing work to tie together the new physics in a holistic synthesis? Biographer Richard Westfall has concluded that in all likelihood there would not. Newton's achievements would no doubt have trickled out eventually or been discovered by others, but the history of physics might well have been quite different, for no one before or since has left such an indelible stamp on the science. Ironically then, at this seminal time the history of physics hinged on allowing a man to remain true to a heretical belief.

But for all Newton's private heresies, publicly he committed his science to the service of orthodoxy, and Newtonian natural philosophy quickly became the basis for a powerful new synthesis of science and Christianity, particularly in the Anglican denomination. Although science was increasingly conducted in a secular environment, under Newton's influence mathematical men continued to serve as a secular wing of the clergy. Following his lead, generations of Newtonians used their science as a powerful tool to defend orthodox Protestant positions about God and his relationship with Creation. Whereas Galileo had highlighted the potential for tension between physicists and theologians, Newton had shown the way to a mutually beneficial relationship. Whereas Galileo had been anxious to wrest epistemological power over nature away from the Church, Newton wanted to share that power. By rejecting the atheistic tendencies of Cartesian mechanism, and by using his cosmology as an argument for an active and providential deity, Newton acknowledged a limit to what science alone could explain. For him, *both* science and religion were fundamentally necessary to a full understanding of the world around us. Rarely has the Church found such an influential ally, and rarely has the bond between physics and religion been stronger than in the mind of this immortal Mathematical Man.

6

＃ ⊘ ÷ ≦ ≪ ∞ ＝ ＋ ± ∼ √ ＞

God, Women, and the
New Physics

THE INTERACTION BETWEEN NEWTONIAN SCIENCE AND ANGLICAN
theology in the first half of the eighteenth century took place on
many levels. Until his death in 1727, Newton himself was a force
helping to shape and strengthen this relationship. From our current perspective, one of the more bizarre aspects of that relationship was the degree to which the new physics was called to the
defense of biblical literalism. Here, Newton led the way by arguing
that, in the light of the new science, the story of the six days of
Creation *could* be taken literally. With the immense weight of his
authority, he pointed out that if the earth did not begin rotating
until the third day, then the first two "days" could be as long as
one liked, thereby giving God ample time to do all that the Bible
asserted. In such efforts he was by no means alone. The potential
of the new physics to provide literal corroboration for biblical
events was also quickly grasped by William Whiston, Newton's successor to the Lucasian Chair of Mathematics at Cambridge (the

position now held by Stephen Hawking). Whiston set out to show that a particular comet he had observed could have been responsible for starting the great biblical flood. Using Newton's laws of gravity and motion, he calculated the comet's trajectory back in time and asserted that it would have been in the right place at the right time to trigger the deluge. Far from making religion redundant, the new physics could serve to fuel the flames of even the most literally minded faith.

A further example illustrates dramatically the extent to which religiously minded Englishmen believed Newtonian science could be useful to theology. In 1699 a mathematician named John Craig published a book entitled *Mathematical Principles of Christian Theology*, which in name as well as content was pointedly modeled on Newton's *Mathematical Principles of Natural Philosophy*. Craig argued that just as mathematical principles could illuminate the heavens, they could throw light on Christian theology. Emulating Newton's scientific masterpiece, he began with three simple laws, which bore a striking resemblance to the physicist's own. Craig's first law asserted that "every man endeavors to prolong pleasure in his mind, to increase it, or to preserve it in a state of pleasure." This was intended as a sort of law of inertia of pleasure, and from such laws and definitions Craig derived Newtonian-sounding theorems of moral and theological import.

Historian Derek Gjertsen has pointed out that "behind this apparent parody of Newton" Craig was pursuing "a serious, if misguided point." From a certain passage in the Gospel of Saint Luke, he had inferred that, when Christ returned to earth, there would be at least one believer left to greet him. Given this, Craig was inspired to think that he could use his "Newtonian" theology to establish an upper limit for the Second Coming of the Savior. Craig's argument began with the assertion that belief in historical events diminishes as they recede into the past. In effect, belief was to be treated like gravity, as a kind of "force" that diminishes in strength the farther away in time one moves from its source. With this model in hand, Craig declared that if we could determine the initial degree of belief in Christ held by his disciples, and also the rate at which such beliefs diminish, then using mathematics we should be able to calculate the time at which belief in Christ would have dissipated entirely. That time would mark the outer limit for the Savior's return. "After much complex reasoning and calculation," says Gjertsen, "Craig concluded that, as the probability of the belief in Christ's return would have declined to zero by the year 3150, he must appear before that date."

Most eighteenth-century attempts to use physics in the service of theology were rather less literal. At the forefront of the new union between Newtonian science and Anglican theology were the influential Boyle lecturers, many of whom were Newton's personal friends. The Boyle lectures were endowed by the seventeenth-century scientist Robert Boyle. According to the terms of Boyle's will, the lectures were to be delivered each year by a cleric in defense of Christianity, and starting in the 1690s they became a prestigious platform from which an array of clerics used Newtonian natural philosophy as a basis on which to argue the existence and nature of the deity. According to historian Margaret Jacob, these lectures quickly became "the cornerstone of a liberal, tolerant, and highly philosophical version of Christianity, a natural religion based upon reason and science." The first of the Boyle lecturers was Richard Bentley, and it had been in preparation for his inaugural series that he had written to Newton inquiring about the relevance of the *Principia* to God. Following Newton himself, Bentley took up the theme that the cosmos demonstrated, through its marvelous design and its ongoing need for divine maintenance, the existence of an active and providential deity. Many other Boyle lecturers followed suit.

At the core of such efforts lay the belief that nature itself could be a primary source of revelation about God. During the medieval era, the Scriptures were considered *the* source of revelation, but, starting with Kepler, mathematical men had effectively raised the theological status of nature. With the Newtonians of the early eighteenth century, this trend toward "natural theology" reached its apotheosis. One of the most influential of all natural theologians was Newton's friend and Boyle lecturer Samuel Clarke. Clarke announced at the start of his lectures that he intended to employ "one only method or continued thread of arguing; which I have endeavored should be as near to mathematical, as the nature of such discourse would allow." Emulating Newton, as John Craig had earlier, Clarke also put forward his theological arguments in terms of propositions and theorems. As well as co-opting Newton's methodology, he cited "discoveries in anatomy and physick" as evidence for a deity. Clarke was so enamored of Newtonian science and its discoveries that he declared that Newton had given theologians a more solid basis on which to demonstrate the existence of God than anyone before. Thus, as historian Roger Hahn has noted, in the early eighteenth century, "belief in God was increasingly grounded in evidence supplied by scientific advances."

According to natural theologians, there were basically two ways science could be used to demonstrate the existence of a deity. First, it could illuminate the supposedly purposeful design of nature, thereby pointing to the need for a purposeful Designer. Here, as we have seen, the solar system was the paradigmatic example, but, in addition to the heavens, Newton saw evidence of God's intelligent handiwork everywhere he looked. A classic example was the human eye. Newton refused to believe that such a wonderful contrivance could arise through "blind and fortuitous" causes. Similarly, the bilateral symmetry of animals spoke to him of a deity with a penchant for mathematical order. During the early eighteenth century, arguments for the existence of God based on the supposedly conscious design of natural systems became tremendously popular, and in 1714 Bernard Nieuwentijt presented an encylopedic tome, *The Existence of God, Demonstrated by the Marvels of Nature,* in which he exhaustively detailed myriad examples. It was an old idea, known to the medievals, but the new science, with its sophisticated analytical techniques and its observational tools (notably the microscope and telescope), infused the "argument from design" with new power.

Ironically, the second way science was used to demonstrate the existence of God was through the things it could *not* explain. This led to the so-called God of the gaps arguments. A classic example was Newton's assertion that, because gravity is always an attractive force, the stars would collapse in on themselves unless there was a God to prevent them from doing so. Because the laws of motion and gravity alone seemed unable to explain the stability of the cosmos, Newton filled the "gap" with God. A particularly interesting "God of the gaps" argument was put forward by the Dutch physicist Willem 'sGravesande, who argued that given the fact that more boys than girls are born each year, it was highly improbable that equal numbers of boys and girls would survive to adulthood without the intervention of an omniscient Being. God was thus invoked to explain what the emerging study of probability supposedly could not.

Both "God of the gaps" and design arguments were notable for the fact that they shared epistemological power between science and religion. Science was taken to be the proper means for describing the operations of nature, but its capabilities were thought to be limited, and beyond a certain point, all explanations had to be handed over to a "higher" power. Here, mathematical men and theologians happily shared the epistemological territory of nature. Each side acknowledged the strengths of the other, and each side ultimately benefited from the relationship—for if, in the early eighteenth century, theology needed the support of science, initially Newtonian science also needed the support of theology. Popular histories often give the impression that Newtonian physics was so compelling its success was a foregone conclusion, but this view ignores the fact that very few people at the time were equipped to understand Newton's laws. Like any radically new idea, the Newtonian world picture had to be sold to the public.

That Newtonianism faced a battle for acceptance is seen by the fact that on the Continent Cartesians vehemently rejected the notion of gravity. Indeed, the "value" of Newtonian science was by no means self-evident, and it took a good deal of public relations to instill it in the English psyche. That in fact had been one of the primary aims of the Boyle lecturers. While Newton had lent the prestige of his name to their version of Anglican theology, they in turn brought the stamp of religious respectability to his philosophy of nature. Faith and physics worked hand in glove.

As had the relationship between the French mechanists and Catholicism, that between the English Newtonians and Anglicanism sprung in part from a mutual desire for a stable and lawful society. With the English Civil War and its aftermath, the seventeenth century had been a time of immense social unrest in England. Magic, witchcraft, and sectarian heresy had been accompanied by outright political revolt against the monarchy, during which England's highly stratified society had, for a brief period, been severely challenged. In the wake of this crisis, many Anglican clerics leapt to the defense of the traditional social order. Newtonian natural philosophy, with its lawful universe ruled by an all-powerful but providential God, provided a view of nature that could serve as a model for the kind of society these clerics wanted to champion. Says historian Margaret Jacob: "The cosmic order and design explicated in the *Principia* became, in the hands of Newton's early followers, a natural model for a Christian society, providentially sanctioned and reasonably tolerant of diverse religious beliefs, provided they did not threaten the stability of the polity."

Thus, alongside Newtonian science arose a *social* Newtonianism that envisaged society as a "natural" order ordained by God. Social Newtonians saw the "stability of the polity" as paramount, and to ensure its continuity, Bentley, Clarke, and Whiston took to their pulpits to preach that the "natural" rulers should be allowed their positions and that no one should be "unreasonably solicitous" to change his or her "station in life." The Newtonian society, like the Newtonian cosmos, was a lawful, stable, immutable, and supposedly God-given order. Just as the planets remained fixed in their respective orbits, human beings were to remain fixed in their respective "stations." It is notable that these Newtonian clerics preached not to the lower classes (who had been the source of much of the unrest in the previous century) but primarily to a prosperous, London-based audience. According to Jacob, they assured their well-to-do congregations that "reasonable people must acknowledge a vast cosmic order, imposed by God, and attempt to imitate it in society and government." It was thus humanity's moral duty to emulate in the social realm the order Newton had discovered in nature. And so, as with French mechanism in the seventeenth century, Newtonian science in the eighteenth was enlisted to justify the status quo. As Jacob and others have suggested, this sociopolitical agenda must be seen as one of the factors contributing to its acceptance.

Part of the "natural," or God-given order that social Newtonianism sought to preserve was the gender order. If men could be viewed as planets fixed in their respective orbits, or stations, then women could be viewed as moons "naturally" compelled to stay in orbit around their men. One station to which women during the eighteenth century were *not* supposed to aspire was formal participation in science itself. During this century, not only did the Royal Society not accept any women members, but not a single woman was granted a degree from any English university. Although none of the major continental scientific societies accepted women as full members either, in Italy a number of provincial societies did accept women, and in Germany a few women were accepted as associate members to the Berlin Academy of Sciences. Similarly, in Italy and Germany a handful of women did gain academic qualifications. In Italy, several women even taught math and physics at universities. Formally, women in France fared little better than in England, but France also had a vibrant salon scene that was an important, if informal part of the emerging culture of science, and within that arena women played a central role in the dissemination of the new philosophy of nature.

The point is that at this time England was both the leading nation in science and the leading nation in denying women a place within the formal community of science. While social Newtonianism was by no means the only factor impeding the progress of women scientists in eighteenth-century England, the strong link between Newtonianism and religion in that country helped to perpetuate a "priestly" view of the scientist—an attitude that in turn helped to perpetuate a climate of thought particularly unwelcoming to women. By stressing the religious associations of physics, one of the legacies of Newton's synthesis was to reinforce the old perception of mathematical science as a sacred activity—a perception that, as we have seen, had long acted as a barrier to mathematical women.

Although in France the new physics had also emerged in a strongly religious climate, by the end of the seventeenth century the priestly undercurrents in scientific culture there were being diluted by the realities of French society. In particular, because the stream of science in France flowed through the salons, there was at least one arena in which women could participate in the scientific enterprise. During the seventeenth and early eighteenth centuries, the French salons were foci of immense social, political and cultural power, and the great *salonnières* such as Madame de Lambert and Madame de Tencin acted as power brokers for ambitious and talented men. Far more than hostesses, these women could make and break careers. It was said, for instance, that if one wished to gain entry to the Académie Française, one must first pass through the salon of Madame de Lambert. As did other ambitious men, many French scientists attended salons and took advantage of the unique combination of society, politics, and power to be found therein.

The French salons were not merely places where scientists whiled away their leisure hours; as historian Mary Terrall has shown, they played a central role in the legitimation of the new science. Above all, notes Terrall, "representatives of science needed to convince their public of the value of mathematics and experiment"; and in France an important part of their public relations strategy was to win the allegiance of powerful salon women. Spearheading this effort during the late seventeenth and early eighteenth centuries was Bernard de Fontenelle, lifetime secretary of the Royal Academy of Sciences in Paris.

One of Fontenelle's strategies was to present science in a form that was both accessible and entertaining to a nonspecialist audience—an audience that he specifically envisaged as including women. In his highly successful book *Conversation on the Plurality of Worlds* (1686), Fontenelle "promoted the ideal of accessible rationality through the fictional device of a naive but intellectually receptive marquise," to quote Terrall. The *Conversation* was framed as an intellectual seduction in which the narrator, a mathematician, "wooed" a sophisticated young marquise with charming and witty conversation about the new cosmology. Other authors soon followed suit. One of the most successful was Francesco Algarotti, who in 1737 penned his *Newtonianism for Ladies*. But if Fontenelle regarded aristocratic women as potential allies of the new science, "at the same time," says Terrall, "he carefully defined academic practice in opposition to salon conversation." Although he co-opted women's help in disseminating and legitimizing the new science, he had no intention of inviting them into the official arena in which scientific knowledge was actually *produced*. Women could be spectators and promulgators of science, but its practice was to remain the prerogative of men. In other words, academicians could go to the salons, but Fontenelle made it clear that salon women could not go to the academy.

In *Conversation on the Plurality of Worlds*, he stressed that the marquise could only go so far in her understanding and that much would remain beyond her feminine grasp. These "higher" truths were reserved for the noble male minds of the academy. Thus, even in France, science retained a "priestly" core and its official locus (the academy) remained strictly male. Yet there was one real marquise who was not content to remain merely a spectator to the new science: Émilie du Châtelet.

Gabrielle-Émilie Le Tonnelier de Breteuil (1706–1749) was nineteen when she married Forent-Claude, the marquis du Châtelet, but, after bearing him three children, the young marquise's interests increasingly turned toward science, and she began to study the new philosophy. At the same time she had become the intimate companion of Voltaire, who had himself recently become fascinated by Newtonian philosophy. Through his connections Voltaire introduced Châtelet to several of the early French Newtonians, including the physicist Pierre Maupertuis, who as a personal favor agreed to tutor her in mathematics. Châtelet proved to be such an able student that when Voltaire wrote his own book on Newtonian philosophy, she supplied the mathematical expertise he lacked, a contribution he acknowledged when he dedicated the book to her.

In 1738, Châtelet began working in secret on her own book about Newtonian physics, but after being introduced to Leibnizian metaphysics by another tutor, she rewrote the entire work. As the book was nearing completion, however, the tutor, Samuel König, happened to see some pages and, considering it beneath his dignity to be known merely as a lady's tutor, put around the story that he was the true author. As Châtelet rightly feared, it was a story male scientists were all too willing to believe and it ruined her chances of a fair reception for her work. Eventually she published the book anonymously. Despite this blow Châtelet remained determined to introduce Newton to the French, and turned to the work for which she is still remembered—a translation of, and commentary on, the *Principia*. To this day it remains the only French translation of that seminal work, and has thus served to introduce generations of Frenchmen to the English physicist. Given the technical difficulty of Newton's text, Châtelet provided assistance for her countrymen by adding extensive explanatory annotations and notes. Unfortunately, just she was completing this massive undertaking, Châtelet discovered that, at the age of forty-two, she was pregnant, and, sadly, she died a few days after the child was born—not living to see her work published.

As had Margaret Cavendish, Émilie du Châtelet tried to enter the discourse about the new science, and as had her predecessor, she learned how difficult it was for a woman to be taken seriously by her male peers. Châtelet too jumped at any opportunity to engage in debate about her work with academicians—even those who criticized her for daring to tread on their turf—but in the male-dominated world of the academy, a woman was, by definition, an anomaly, and Émilie du Châtelet always remained an outsider.

Châtelet was not the only continental woman of the early eighteenth century to make an important scientific text available in her own language. In 1722 Giuseppa Eleonore Barbapiccòla published an Italian translation of Descartes' *Principles of Philosophy*, a work she carried out while still in her late teens or early twenties. Noting that the philosopher had dedicated this text to Elizabeth of Bohemia, Barbapiccola argued that Descartes had intended *women* as his principal audience, and they were the audience she herself aimed to reach. In her preface she wrote: "I yearned to translate it into Italian to make it accessible to many others, particularly women who, as the same René says in one of his letters, are more apt at philosophy than men."

Later in the century, Maria Angela Ardinghelli (1728–1825)

was responsible for introducing Italy to the work of Stephen Hales, perhaps the most important early Newtonian scientist after Newton himself. Ardinghelli not only translated Hales' words, but recalculated all his results, corrected his mistakes, and redid experiments that seemed unclear to her. According to historian Paula Findlen, she "did not neglect to demonstrate her superior mathematical abilities by clarifying his work at every stage." Unlike Châtelet, who had always regretted the crude state of her education, Ardinghelli had been tutored in mathematics and physics from childhood, and by the age of twenty was amazing Neapolitan salons with her knowledge of electricity—the latest scientific vogue. Ardinghelli was introduced to the French physicist-priest the Abbé Nollet, who soon became a mentor, and over the years she corresponded with him on a wide variety of scientific subjects. As a member of the Paris Academy of Sciences, Nollet encouraged fellow members to correspond with her, but, as he regretted, she herself could not become a member of their august institution. Unable to change the admission policy even for this demonstrably capable woman, the enlightened Abbé convinced the academy to request her portrait for their gallery. An *image* of a woman was apparently the most these men could tolerate in their midst.

In the eighteenth century, there was, however, one Italian woman who crashed the institutional ramparts of science and became an accepted participator within that community. Yet in her very success Laura Bassi (1711–1778) reveals the limits of what a woman physicist could achieve at this time. As did all our mathematical women so far, Bassi received her education at home. The daughter of a Bolognan lawyer, she was schooled by the family physician, Gaetano Tacconi, and by the age of twenty, her fluency in Cartesian and Newtonian natural philosophy was beginning to attract attention. She was reputed to be a "monster in philosophy." In the face of mounting pressure for public displays of her erudition, Tacconi allowed a select group of professors and learned gentlemen to hear Bassi dispute a variety of topics. They were so impressed that in the following months she was elected to the Academy of the Bologna Institute of Sciences, and also invited to be examined for a degree at the University of Bologna. Passing this test with ease, on May 12, 1732, she became only the second woman *ever* to gain an academic qualification. The first was the Venetian noblewoman Elena Cornaro Piscopia, who in 1678 had been awarded a degree in philosophy from the University of Padua. Finally, in an unprecedented move, the senate of the University of Bologna offered Bassi a chair. Thus, in 1732 she became the world's first woman professor.

Yet the young "philosophess" was by no means just another university teacher, for the senate had appointed her under stringent conditions. She was only to lecture when *they* determined, and in practice this turned out to be only at special public occasions. In effect, says Paula Findlen, the university intended her to be an exotic ornament to be put on display to attract fame and attention to this once great but now rather faded institution. As a public relations exercise, Bassi was an immense success: People flocked to hear her lectures, and visiting scholars from all over Europe stopped by to see for themselves the famed *doctoressa*.

But Bassi had ideas of her own, and not content to be merely an academic bauble, she began to expand her role. An early hint of her independent spirit came in her decision to marry. In an age when women were supposed to do little else, a decision to marry may not sound like a radical act, but the senate had conceived of Bassi as some kind of symbolic learned virgin. In this expectation, says Findlen, she recalled "both the religious tradition of women in orders as brides of Christ and the civic tradition of virgins whose chastity cemented the foundations of republican government." During her degree ceremony, Bassi had received a ring signifying her virtual marriage to both the city and university. Bassi refused to conform to this expectation and in 1738 she married the physicist Giovanni Giuseppe Veratti, to whom she bore eight children.

Being a mother raising a family did not diminish her determination to pursue her physics, and she convinced the university to expand her schedule of public lectures. But there was a limit to how far they were prepared to let her go, and Bassi never became a regular academic. Her role at the university remained strongly ceremonial, though by 1760, says Findlen, she earned "a salary higher than that of any of the other professors and members of the Institute, including the president."

Again, not content to be restricted to the activities approved for her by the university senate, from 1749 Bassi offered private lessons in experimental physics at her home and was a champion of Newtonian physics when it was not yet widely known in Italy. She also began to collaborate with her husband on electrical research. Throughout her life, Bassi remained a prominent member of the institute, producing scientific papers for their meetings and managing their famous school for experimental physics. She also corresponded with a number of the greatest physicists of the day, including Roger Boscovitch (the founder of the idea of a universal force of nature) and Alessandro Volta (the pioneer of electric research). Finally, at the age of sixty-five, Bassi was appointed to the institute chair of experimental physics. This time it was the husband who became the assistant!

What Bassi achieved was truly remarkable. Yet, despite the fact that she played a major role in introducing Newtonian natural philosophy and experimental physics to Italy, she has always been known, says Findlen, "for her exceptional circumstances, rather than her ideas." If she was privileged in her unique role as a woman inside the institutional landscape of eighteenth-century science, she was also imprisoned by it. As Findlen has stressed, she constantly strove to "regularize" her position, but she could never simply be a physicist, she was always *the woman physicist,* whose every move was a topic of discussion and debate. Although she was accepted and respected in a way that Châtelet could only dream of, Bassi remained an anomaly in a male world. Furthermore, for all her personal success, Bassi did not succeed in opening up the path for other women. Says Findlen: "Having discovered how far an ambitious and persistent woman like Bassi could insinuate herself within the Institute, the male members of its academy were noticeably reluctant to allow another woman such latitude." Not until Marie Curie, more than a century later, would any other woman establish herself so strongly within the male-dominated culture of physics.

As we have seen, throughout history mathematical women have relied on the help of enlightened men for access to education and opportunity, and Bassi's case was no different. Her mentors included her father, who ensured she received a scientific education in the first place; Tacconi, who tutored her; and her husband, who supported her unorthodox choices and who lobbied for *her* to receive the institute chair in experimental physics, rather than himself. But the most important of Bassi's champions was Prospero Lambertini—cardinal, archbishop, and eventually Pope Benedict XIV. As a Bolognan native, Lambertini was anxious to help resurrect the scientific glory of his beloved city, and it was he who perceived what a public asset Bassi could be. From the beginning he was an important patron, and it was rumored that he had ordered the university to grant her a degree.

That a pope, no less, could champion a woman in science is evidence that the Church need not necessarily foster a misogynist culture of learning. Reinforcing this view is the fact that Maria Ardinghelli's mentor was the Abbé Nollet, who also corresponded with Bassi. Lambertini may also have had ulterior motives for supporting Bassi, but his feminist stance is no less admirable for that. Under his influence Bologna became a unique (albeit limited) haven of opportunity for women in science. The institute's academy asked not only Bassi to be a member, but also Émilie du Châtelet. Similarly, in 1751 the University of Bologna awarded a degree to another woman—the prodigy Cristina Roccati—making her only the third woman *ever* to gain an academic qualification. Roccati (1732–1797) was on her way to becoming another Bassi when her family was financially ruined and she was forced to retire to provincial Rovigo. Lambertini also tried to woo to Bologna the mathematician Maria Gaetana Agnesi (1718–1799). In this case, it was she who did not wish to avail herself of the rare opportunity.

Perhaps more poignantly than anyone, Maria Agnesi highlights for us the limitations on women in science during the eighteenth century. Like Hypatia, she was the daughter of a mathematician who educated her to the highest standards, but unlike Bassi and Châtelet, Agnesi disliked the social whirl of the salon scene and longed for a quiet religious life. After writing two books—one on the new natural philosophy, and the other a brilliant text on differential and integral calculus, which synthesized the mathematics of Descartes, Newton, and Leibniz—she withdrew from society and spent the rest of her days living devoutly, dissipating her inheritance on charitable works. Whatever a woman's inclinations, she was still expected to be a society lady—and that was a role Agnesi was not prepared to play.

No woman of the eighteenth century had the option to live like Newton—an antisocial loner devoted to her scientific work. If Agnesi had been a man, she could have pursued mathematics while being in a religious order, just as the Abbé Nollet pursued his physics. But as a woman, she had no such option available to her, and was forced to choose between the two ideals. Who knows what she might have achieved had the possibility of mathematician-monk been open to her? This is a question we may well ask with respect to all the women in this chapter: What might any of them have achieved had they had the opportunities their male peers took for granted?

However meager the opportunities for mathematical women during the early eighteenth century, during the latter part of the century their options actually declined. Ironically, this downturn was coincident with a major parting of the ways between physics and religion. Whereas early Newtonians had happily used their science in the service of theology, as the century wore on (and Newtonianism swept across Europe), mathematical men sought to wrest their science away from its traditional religious moorings. Yet despite the increasingly secular climate in which physics was practiced, mathematical men of the Enlightenment retained a quasi-religious attitude to their own activities, and along with this the majority retained a quasi-clerical attitude about who was fit to participate. In particular, they continued to stress that physics was rightfully a male-only enterprise. Rather than being at odds with Enlightenment rationalism, this old impulse gained fresh vigor from the new emerging philosophy.

The late eighteenth century, with its social rationalism, its republicanism, and its anticlericism, saw a distinctive change in heart among mathematical men, particularly in France. Whereas Newton and his early English followers had been anxious to share epistemological power over nature with theologians, as the century progressed physicists increasingly wished to claim the whole of nature for themselves. They began to believe that nature was self-sufficient and did not require the hand of God to keep it functioning properly. For instance, instead of attributing gravity to a higher power, as Newton had done, physicists now saw it as simply an innate property of matter. Furthermore, many of the things that Newton had believed science could not explain, and which he had therefore attributed to a Supreme Being, *were* gradually yielding to physicists' analysis.

This trend reached a climax in the work of the French physicist Pierre-Simon Laplace (1749–1827). In opposition to Newton, by the mid-eighteenth century, most astronomers had come to believe that the solar system would run forever without divine intervention. The problem with this belief, however, was that astronomical evidence did not seem to support it. Observations showed that the planet Jupiter was slowly accelerating and Saturn slowly decelerating; in the long run, such a system could not be stable. Yet the tenor of scientific thinking had changed so much that, rather than see this anomaly as evidence for a God who must step in and reset the cosmic mechanism from time to time, astronomers now believed the fault must lie in their own methods of calculation. The "gap" was not in the cosmos, they believed, but merely in their understanding.

Confirmation of the self-sufficiency of the solar system was at last provided by Laplace, who in 1786 presented to the French Academy of Sciences a mathematical demonstration that the aberrant motions of Jupiter and Saturn would be reversed in a few hundred years. Laplace showed that these anomalies form a cycle, wherein one planet speeds up while the other slows down, then vice versa. Thus the solar system was naturally stable, and Newton's laws alone were sufficient to account for this.

Belief in the self-sufficiency of nature gave physicists a basis from which to argue for the separation of science and theology. By showing that the solar system was self-sustaining, Laplace not only obviated the need for a higher power to perform this function, he also negated one argument for God's very existence. Increasingly, physicists such as Pierre Maupertuis and mathematicians such as Jean Le Rond d'Alembert began to argue against natural theology, and to insist that the material world could not be used as evidence for the existence of a deity. A leading proponent of this view was the German philosopher and physicist Immanuel Kant (1724–1804). According to Kant, gaps in present scientific understanding could never be taken as proof of God but must instead be left for the science of the future to explain. In essence, Kant declared that "God of the gaps" arguments were invalid. Neither would he accept the argument from design. Kant called into question the whole project of natural theology and thereby undermined the relationship between science and theology that early Newtonians had established. He did not deny the existence of God (Kant was in fact a devout believer), but he denied that science could serve theology.

More radical even than denying the notion of God as active overseer of the universe, Kant and Laplace began to challenge his role as Creator. In 1796 Laplace put forward a hypothesis about how the solar system could have formed by natural processes alone. The details of this hypothesis are not important; what matters is that mathematical men were beginning to suggest that the genesis of the heavens could be explained by science rather than religion. Kant went even further. Following Descartes, he believed the entire universe could have come into being through mechanical processes. In effect, he said, nature made itself, and God was redundant in the process of cosmic creation.

A public endorsement of this view appeared in the *Encyclopédie Méthodique,* in which an article titled "Order of the Universe" boldly declared that God soon would be exposed as an "excess wheel in the mechanism of the world." But the most famous statement of God's irrelevance to Creation came from Laplace himself. In 1802 the French physicist was conversing with Napoleon when the talk turned to the new celestial mechanics and Napoleon innocently asked who the author of this marvelous system of the heavens might be. According to legend, Laplace is said to have replied that, with respect to God: "I have no need for that hypothesis." For once in his life, Napoleon was supposedly speechless.

By writing God out of nature, Laplace and Kant were claiming it wholly for science. Whereas Galileo had tried to separate the planetary system from religion, Laplace and Kant now wanted to claim the entire physical universe, including its creation. In doing so, they consciously pitted themselves against theologians, for one could not lay claim to cosmic genesis without treading on theological territory. Again, this was not simply a matter of pitting reason against faith, for just as the science of Galileo's day could not prove that the earth moved, the science of Laplace's day was not equipped to demonstrate that the solar system or the universe evolved by natural processes. Laplace and Kant's claim was based not on scientific evidence but on another kind of faith—faith that science *would* ultimately be able to explain all of nature.

Much of that faith stemmed from advances in techniques of mathematical analysis made during the eighteenth century. Just as Laplace had demonstrated in a mathematical tour de force that the solar system was self-sustaining, confidence was growing that mathematical analysis would ultimately be able to reveal all of nature's innermost secrets. During this century mathematicians such as Joseph-Louis Lagrange had developed powerful techniques that enabled Newton's laws to be applied to a vast range of situations, including the motion of fluids and the behavior of nonrigid bodies. Through the prism of mathematics, physicists were gaining insight into light, heat, sound, fluid flow, elasticity, and electricity. All in all, mathematics was proving to be an immensely powerful tool for penetrating the hidden order beneath the veil of the senses. Was there anything it would not be able to elucidate?

According to Laplace, mathematical analysis could lead human beings to an almost godlike omniscience. In a famous statement, he suggested that if there was "an Intelligence" that knew the exact state of every particle in the universe at one instant, then, knowing the laws of nature, this Intelligence would be able to calculate the exact state of the universe at any moment in the past or future. As Roger Hahn has pointed out, Laplace's Intelligence can be seen as a vestigial God. Here was the old Christo-

Pythagorean deity stripped of all moral, redemptive, and providential function—a god merely as mathematical calculator. What is most startling about Laplace's "god" is how much it resembled a super version of the physicist himself. Whereas Laplace could personally calculate the state of the planets, the Intelligence could calculate the state of the entire universe. Yet if Laplace's god was a kind of superphysicist, then reciprocally the physicist became a kind of earthly god—or at least he could aspire to be. Thus, while Laplace may have wished to sever his science from theology, he had no intention of diminishing physicists' claim to "higher" truth. While distancing physics from institutional Christianity, he still wished to retain a Pythagorean religious sentiment toward the actual science. And, despite the increasingly secular climate, physicists of the late eighteenth century were seen to be engaged in a quest for quasi-divine knowledge.

That attitude is evident in what can only be called the deification of Newton. The process had already begun during Newton's lifetime, when in his introduction to the *Principia*, Edmund Halley had declared: "Nearer the gods no mortal may approach." Taking that sentiment further, Alexander Pope penned the well-known epitaph:

> Nature, and Nature's Laws lay hid in Night.
> God said, *Let Newton be!* and All was Light.

Seven years later, in 1735, the British government built its Temple of Worthies as a monument to national genius, and the inscription over Newton's bust began with the words: "Sir Isaac Newton, Whom the God of Nature made to comprehend his works: and from simple Principles, to discover the laws never known before."

In the latter part of the eighteenth century, the French would go to even greater extremes to deify Newton. Architect Étienne-Louis Boullée designed a huge cenotaph to the English physicist. Although never built, it was to carry the legend "Sublime spirit! Vast and profound genius! You are divine!" Meanwhile, Champlain de la Blancherie denounced the English for failing to honor Newton's divinity sufficiently. The proper homage, Blancherie felt, was to restart the numbering of the years from 1642, the year of Newton's birth. Newton would replace Christ as the origin of the Western calendar.

One by-product of this quasi-religious attitude toward physics was to fan the flames of the old belief that the quest for nature's "laws" was a "priestly" activity suitable for men alone. Kant (who trained as a theologian, and avoided the company of women throughout his life) declared that science properly has a "masculine mien" and railed against women who dared to venture onto this hallowed ground: "A woman who has a head full of Greek, like Madame Dacier, or one who engages in debate about the intricacies of mechanics, like the marquise du Châtelet, might just as well have a beard; for that expresses in a more recognizable form the profundity for which she strives." Elsewhere he wrote: "The fair sex can leave Descartes's vortices to whirl forever without troubling itself about them." Again, the absence of women from physics in the late eighteenth century cannot be blamed entirely on the religious undercurrents of the science, for Enlightenment sentiment against educated women in general was endemic. In 1788, for instance, Christoph Meiners, a professor at the University of Göttingen, began a four-volume history that he hoped would deliver Europe from "the calamity of pedantic women." Nonetheless, the covert religiosity underlying physics continued to act as a powerful additional barrier to women.

In the middle of the eighteenth century those who would have science retain a "masculine mien" increasingly began to attack the entire salon scene, the one arena in which women could freely mix with men of science. According to Jean-Jacques Rousseau, the company of women lowered the level of discourse among men and thereby led to a decaying of the nation's intellectual fiber. In the presence of women, Rousseau said, men are forced to "clothe reason in gallantry," but by themselves they engage in more "grave and serious discourse." As an alternative to salon gallantry, Rousseau suggested a military metaphor for learned discourse, one in which men engaged in disputes, defending their ideas as if on an intellectual battlefield. He drew a parallel between a strong mind and a strong body, and asserted that women lacked the strength to participate in science. For serious intellectual engagement, Rousseau advocated that men must withdraw to male-only clubs or "circles."

For Rousseau, the salons also encouraged an unacceptably poetic and "feminine" style of scientific writing. Poetry, dialogue, and other literary devices had in fact always been employed in scientific works, and many *salonnières* proudly adopted a graceful style. Men such as Denis Diderot, the editor of the great *Encyclopédie*, and the comte de Buffon, the French naturalist, prided themselves on turning out an elegant phrase. Both Diderot and Buffon believed the company of salon women had a beneficial effect on the male intellect, forcing men to clarify and refine their ideas. "Women," said Diderot, "accustom us to discuss with charm and clearness the driest and thorniest subjects." But the poetic style associated with the salons was increasingly decried, and opponents championed a supposedly more "masculine" prose, stripped of literary embellishment.

An insight into the changing style of scientific writing can be gained by comparing Buffon with Laplace. Several decades before Laplace put forward his nebula hypothesis for the genesis of the solar system, Buffon had advanced a similar idea in his bestselling *Natural History*. Buffon's book has been described as "a cosmological novel," but when Laplace took up the idea at the end of the century, he relied heavily on the new mathematics of probability. By the middle of the next century, says historian Londa Schiebinger, poetry had largely been eliminated from official scientific writing as it became ever more abstract, mathematical, and technical. This new trend disadvantaged all nonspecialists, but it particularly disadvantaged women, who were still entirely cut off from the channels of higher education.

As Schiebinger has stressed, however, there is nothing innately feminine about a poetic style of writing; in other ages poetry has been seen as distinctly masculine. The issue here was that science was being increasingly defined in opposition to whatever was seen to be associated with women, and, reciprocally, women were being associated with the things cast out of science. The very concepts of "science" and "femininity" were being constructed as *opposites*.

Whereas science was associated with reason, objectivity, and facts, women were associated with feelings, subjectivity, and literary allusion.

Enlightenment construction of women in opposition to science had its basis in the theory of complementarity, which held that women and men were two different but complementary kinds of beings. Complementarians believed that male and female were made by nature for entirely different purposes, each being necessary to the functioning of the social whole. In theory the two sexes were equal, but in practice difference meant hierarchy because all the qualities most highly valued were those supposedly possessed by men, while those least valued were associated with women. In particular, complementarians placed the kind of reasoning needed to do science on the slate of male attributes. According to Londa Schiebinger, complementarians taught that "women lack the genius to engage in the search for abstract and speculative truths." Indeed, Rousseau, Kant, Meiners, and a long list of Enlightenment intellectuals "all held that creative work in the sciences lay beyond the natural capacities of women."

Soon science itself was enlisted to the complementarian cause as practitioners in the emerging field of anatomy searched for scientific evidence of women's intellectual inferiority. After careful measurement, anatomists "discovered" that women's skulls were smaller in proportion to their bodies than men's. Thus, they said, the facts demonstrated that, as thinking beings, women were inferior to men. The problem with this deduction was that women's heads are actually *larger* in proportion to their bodies than men's. When anatomists were forced to concede this point in the nineteenth century, they did not thereby conclude that women had better brains; instead they interpreted the relatively larger head as a sign of incomplete growth. Cranial size was seen to indicate that women were closer to children, whose heads are also proportionally larger. Thus, again, women were construed as mentally inferior to men, and so, unsuited to the pursuit of science.

The eighteenth-century feminist Mary Wollstonecraft (mother of Mary Wollstonecraft Shelley, author of *Frankenstein)* pointed out that male scientists were *defining,* rather than *discovering,* women's nature as incapable of doing science; and feminists, male and female, rejected this blatantly self-serving process. Theodor von Hippel, mayor of Königsberg and an acquaintance of Kant, denied that anatomists' discoveries had any significance for the mind. Whatever differences had been found, said Hippel, could not be equated with intellectual capacity. Yet champions of women were no match for the authority of science, which together with the new Enlightenment philosophy created a powerful basis for the exclusion of women not only from science, but from the polity at large.

Although this was the era when France rang with cries of *"Liberté!"* and *"Égalité!"* the most germane part of the triumvirate was *"Fraternité!"* In the wake of the French Revolution, the republican government denied political rights to women, citing newly established definitions of women's nature as evidence that they did not possess "the moral and physical strength" necessary to exercise the right of citizenship. As historian Joan Landes has pointed out, the new conception of "the people" recognized men only, and, for the first time in history, women's disenfranchisement was enshrined in law. Rather than being encouraged to enter the public sphere, women of the late Enlightenment were told that their proper sphere of action was the private realm of the home. Philosophers such as Rousseau, Kant, and Georg Wilhelm Friedrich Hegel justified this division of society by appealing to nature and its laws. Their view of the social order was based on a conception of an immutable "natural" order, in which women were grounded in the immediate, the private, and the familial, whereas men were seen as naturally suited to universal concerns—especially the public law of the state.

This public-private, male-female dualism provided fresh impetus for the old clerical prejudice that the study of nature's laws was a task for men alone. If women were "naturally" situated in the private sphere and men "naturally" dominated the public sphere, then the "laws of nature" naturally fell into men's territory—for they are the most public laws of all. Thus, rather than being a force for increased equity between the sexes, Enlightenment philosophy reinforced the old Aristotelian mentality, thereby ensuring that in the "modern" age women continued to be excluded from the official arenas of science. When Rousseau wrote to d'Alembert that women lack the "celestial flame," he may as well have been speaking from the thirteenth century. As before, nature was to remain the provenance of Mathematical *Man*.

In the wake of the French Revolution, the salons disappeared, and with them, the one arena in which women could mix easily with men of science. In the *ancien régime,* at least some noblewomen had been able to exchange access to rank for access to scientific knowledge; but in the new climate of "equality," *all* women were equally excluded. The opportunities that had begun to open up for mathematical women in the early part of the century were shut off and women once again found themselves in an almost medieval relationship to the official world of science. As a letter to the editor of a contemporary magazine lamented in 1796: "In all ages men have sought to distance women from knowledge; but today this opinion has become more fashionable than ever."

7

$\Omega \quad \mu \quad \delta \quad \Omega \quad \omega \quad \pi \quad \Sigma \quad \Delta \quad \gamma \quad \beta \quad \alpha \quad \text{℔}$

Science as Salvation

WHEN ONE CONSIDERS THE BITTER WARS THAT HAVE BEEN FOUGHT between religions or between different branches of the same religion, the relationship between Christianity and science in the seventeenth and eighteenth centuries was remarkably harmonious. It could be so because, in ceding nature to scientists, theologians were not in fact giving up anything central to Christian faith. The core of Christianity had never been its conception of the material world, but rather its spiritual dimension—in particular its promise that, through Christ, all human beings, regardless of class, race, or gender, could be redeemed into eternal grace. Even after physicists began to claim *all* of nature for science, the spiritual core of Christianity remained intact. The separation of the material and the spiritual into distinct disciplines certainly constituted a break from the unity that had characterized earlier Christianity, but it was by no means an unworkable partnership.

As long as the practitioners of science confined themselves to the elucidation of the material world, there was no reason that science and religion could not have continued to coexist happily. One of the great strengths of Christianity has been its ability to ultimately absorb and harmonize scientific discoveries—a process that continues today. In recent years, for example, the Vatican has been co-hosting a series of conferences on science and religion with the San Francisco–based Center for Theology and the Natural Sciences. At these conferences, theologians, scientists, and philosophers come together to discuss the interaction of science and faith in the contemporary world.

What Christianity has not been able to withstand, and what is the source of a good deal of the tension between science and religion in our own century, has been the challenge to its *spiritual* core—the challenge to the idea that Christ is the one true path to salvation. While at first glance it might seem that the redemptive role of Christianity would be immune to incursions from science, in fact, as early as the seventeenth century, science was touted as an indispensable aid to Christian salvation. Recall that Newton aimed to restore the "true" Christianity by the restitution of the "true" scientific knowledge of the world. In the nineteenth century, however, champions of science began to present it not merely as an aid to religious salvation but as the path to human salvation in and of itself. This new trend arose from the growing belief that the technology now emerging out of science would create a "heaven" here on earth. A key feature of the century following the Enlightenment was the rise of the idea that humanity would be saved not by religion but by science and its technological by-products. At the forefront of this movement was Mathematical Man, for it was from physics that many of the new technologies of the nineteenth century would come.

Although the idea of salvation through science was not fully manifest until the late nineteenth century, its origin can be traced to the English philosopher Francis Bacon, a contemporary of Kepler and Galileo. While no great theories attach to Bacon's name, no one has done more to shape the ethos of modern science. When the Royal Society was formed in Britain in 1660, it was widely regarded as the realization of Bacon's vision of a scientific community, and his dream of what science might be has inspired its practitioners ever since.

From the beginning Bacon (1561–1626) seemed destined for a brilliant career. His father had risen from humble origins to become lord keeper of the great seal at the court of Queen Elizabeth I, and his mother, herself a well-educated woman, had recognized early on her son's precocious intellectual gifts. She subjected him from a young age to a rigorous regimen of ancient and modern authors, and this Renaissance enrichment program led Francis to Cambridge University at the tender age of twelve. At fifteen he left Cambridge with a reputation for prodigious application and entered into training as a lawyer—not with the intention of going to the bar, but as preparation for a career in state administration. After having risen to the ranks of solicitor general, judge of the court of the verge, then attorney general, he too was appointed lord keeper of the great seal, and finally lord chancellor. Bacon's contributions to the state were recognized when he was knighted by James I, successor to Elizabeth, and later granted the titles Baron Verulam and Viscount St. Albans.

In spite of these honors, Bacon's career ended in shame in 1621, when he was accused of taking bribes from criminals over whose trials he had presided. While he readily admitted to taking the money, Bacon denied that his judgment had been influenced by the spoils. But therein lay his mistake, for what he failed to appreciate was that the usual exchange for a bribe is acquittal. Bacon's misdeeds had come to light only because several criminals complained that he had taken their "gifts" and then gone on to convict them. It is simply staggering that one of the greatest legal minds of the century did not understand that if one cannot stoop to tipping the scales of justice, one can hardly expect to take a bribe and get away with it. But if Bacon's legal career came to an end, his public downfall freed him to pursue his private passions. Banned from holding public office, he pursued a literary career, penning his seminal *Essayes* on subjects such as friendship, love, ambition, atheism, and truth. Above all, he threw himself into his great project of developing a new philosophy of science. Along with Kepler in Germany and Galileo in Italy, Bacon was at the forefront of those calling for a new approach to nature.

As would Mersenne and Descartes, Bacon stressed that science could serve faith. In particular he believed it could contribute to human salvation. Whereas Newton would be obsessed with restoring to humanity the original Adamic knowledge of nature, Bacon was obsessed with restoring what he believed was the original Adamic power over nature. Through science he believed humanity could win back the state of grace—and the accompanying power— it had possessed in the Garden of Eden before the Fall. The goal of science, Bacon declared, was nothing less than "the restitution and reinvesting of man to the sovereignty and power . . . which he had in his first state of creation." The faith that science could restore humanity to a prelapsarian state of grace was based on the belief that from science would flow all sorts of technologies that would improve the material conditions of people's lives. Science would "save" us by giving us the power to make all lives happy, healthy, and comfortable—in short, more Edenic. If man was the subject of power through science, then its object was nature. According to Bacon, science would give us "the power to conquer and subdue" nature. Elsewhere he announced: "I am come in very truth leading you to Nature with all her children to bind her to your services and make her your slave."

Bacon outlined his vision of the world science would bring about in his famous treatise *The New Atlantis* (1627), a short tale that reads rather like a chapter out of *Gulliver's Travels*. As with Swift's classic, Bacon's tale is told by a narrator who describes how his ship was driven by a violent storm onto the shores of a marvelous land—the New Atlantis. There, the citizens live ideal Christian lives in an ideal society, all made possible by a scientific institute cum monastic colony known as Solomon's House. It was this visionary organization that would so inspire the founders of the Royal Society later in the century. In the New Atlantis, all people live in harmony, without crime or promiscuity; no man robs or assaults his neighbor; and, except within the sacrament of marriage, all are strictly chaste. As the narrator informs us, "there is not under the heavens so chaste a nation as this" nor one "so free from all pollution or foulness."

As an example of the elevated lives of the New Atlanteans, the narrator describes an elaborate ceremony held when any man reaches the point of having thirty living descendants. To such a man the king grants "gift of revenue, and many privileges, exemptions, and points of honor." Among the royal tributes is a bunch of gold grapes. From now on, whenever the father appears in public, this golden cluster will be borne before him by one of his sons as "an ensign of honor." Yet while children, friends, and governor pay homage to the father, the mother is kept hidden behind a screen, "where she sitteth, but is not seen." In the New Atlantis, women not only know their place but are apparently content to be invisible. Not surprisingly, then, they play no role in the true glory of these noble people—the practice of science.

The locus of this defining activity, Solomon's House, resembles nothing so much as a monastic order. The core of the institution is a group of thirty-six "fathers" whose self-appointed mission is to find "the knowledge of causes and secret motions of things" and to use such knowledge for "the enlarging of the bounds of human empire, to the effecting of all things possible." Above all, the fathers of Solomon's House strive to apply their knowledge of nature to the betterment of the citizens of the New Atlantis. And because they are not just scientists, but also priests, their industry leads to a better society morally, as well as materially.

In Solomon's House "all things" do seem to have been made possible. Artificial metals, gems, and stones have been produced, likewise wonderful new kinds of materials and ceramics. All manner of engines and machines have been fabricated: flying machines, submarines, and machines that imitate the motions of living creatures such as birds and fishes. There are also machines of war and new kinds of gunpowder. Alongside these methods of extermination, the fathers have created a huge range of medicines for healing the sick and prolonging life. They have made marvelous optical instruments: spectacles, telescopes, microscopes, and devices that produce illumination. In addition, there are devices to improve hearing and convey sound through pipes over long distances. The fathers have also simulated atmospheric phenomena such as wind, rain, hail, snow, thunder, and lightning; they have even emulated the fire of the sun and stars. And they can predict natural disasters such as earthquakes, floods, diseases, and plagues, and so direct the populace to take precautionary measures.

One astonishing point about all this is how many of Bacon's imaginings have been realized by modern science. Reading the list of the fathers' achievements, one cannot help but be struck by how comprehensive it is: About the only element missing is the computer. One of Bacon's most prescient insights into the direction science might take was his vision of what its practitioners would do with living creatures. The fathers of Solomon's House not only crossbreed animals to produce chimeras; they can make them "greater or taller than their kind is, and contrariwise, dwarf them, and stay their growth . . . also [they] make them different in color, shape, activity—many ways." Similarly, the fathers conduct "trials" on animals for the purpose of learning "what may be wrought upon the body of man"; in particular they try "all poisons, and other medicines upon them." Here then was the first articulation of the idea that animals could be used for testing drugs and other chemicals.

Bacon's vision of science in *The New Atlantis* is amazing if for nothing else than its sheer scale. In order to perform weather simulation experiments the fathers have hollowed out entire mountains and dug vast caverns beneath the earth, "some of them above three miles deep." They have also built enormous towers that ascend into the upper reaches of the atmosphere. Within their laboratories and workshops they are served by "a great number of servants and attendants." Indeed, they seem to have access to unlimited resources. Granted absolute freedom, these men who can imitate the sun are revered as high priests and treated like kings. When any father appears in public, the entire populace turns out to greet him, and he is "carried in a rich chariot . . . all of cedar, gilt, and adorned with crystal, save that the fore-end had panels of sapphires set in borders of gold, and the hinder end the like of emeralds." He is attended by fifty young men dressed in white satin, and before the chariot walk other attendants carrying a pastoral staff and crosier, the symbol of a bishop. Given such reverence and funding, is it any wonder the founders of the Royal Society aspired to emulate this fictional brotherhood?

Yet although the vision of science as a transformative power was present from the early seventeenth century, not until the late nineteenth did that vision bear significant fruit. In the interim, physicists achieved a great deal of theoretical understanding, but little of this knowledge translated into concrete technological advances. Clocks, telescopes, pumps, and mining equipment were all improved, but, by the mid-nineteenth century, the new physics had produced no major technological breakthrough. Modern "science" had changed the Western mind-set, but it had not yet changed the material condition of people's lives. Finally, in the latter half of the century, physics gave rise to a technological revolution that altered the tenor of life in the Western world, and increasingly in the rest of the world as well. Mathematical Man now became not just a Pythagorean being, but also a Baconian one.

This neo-Baconian revolution was ushered in by a new branch of physics known as thermodynamics. As its name implies, thermodynamics is the study of heat flow, and although it remains one of the most unglamorous branches of physics, few branches of any science have had such an impact on daily life, for it was a knowledge of thermodynamics that allowed engineers to design and build efficient steam engines. Originally invented in the early eighteenth century to pump water out of mines, by the early nineteenth century, steam engines were being put to use in flour mills, lumber mills, sugarcane mills, cotton gins, and grain-threshing machines. Thus, the Industrial Revolution began. But these early engines were extremely inefficient and hence not very cost-effective. If industry was going to flourish on steam power, improvements would have to be made.

A breakthrough came in 1824, when a young French engineer, Nicolas-Léonard-Sadi Carnot, outlined a set of scientific principles on which to assess the operating efficiency of heat engines. During the next half-century, the laws of thermodynamics were developed, providing a formalized mathematical basis on which to improve engine design. This theoretical understanding made possible far more powerful, efficient, and compact steam engines. Mathematical Man had thereby provided the foundation for the explosion of mechanization that marked the late Industrial Revolution. Improved engines translated directly into automated farm equipment, steam-powered boats, and better steam trains. But above all, powerful, efficient engines meant that industry could now be conducted on a hitherto undreamed-of scale. Instead of being processed in small local mills, grain and lumber could be brought to centralized plants and processed in bulk. Likewise, secondary and tertiary goods could be mass-produced in huge factories.

The next Baconian breakthrough came from an entirely unexpected quarter. Unlike the founders of thermodynamics, the founders of electromagnetic theory were not trying to solve any practical problems but merely trying to understand the mysterious phenomena of magnetism and electricity. Yet from their endeavors sprang what remains the single most important application of modern science—electric power.

Magnetism and static electric phenomena had been discussed since the Golden Age of Greece, yet in the early nineteenth century they remained enigmas, because they could not easily be fitted into the Newtonian paradigm. The modern, and distinctly un-Newtonian, explanation of these phenomena was suggested by another English genius, Michael Faraday (1791–1867). The Cinderella of science, Faraday was born into poverty and spent much of his childhood hungry. After receiving only a basic education, he was apprenticed to a bookbinder at the age of fourteen, and there he had the opportunity to read the books that came through the workshop. One of the subjects that caught his attention was the new research in electricity. Yet Faraday lived in an age when science was still largely a gentlemen's pursuit, and the opportunities for an uneducated bookbinder were effectively nil. But he was not deterred, and, after attending a series of public lectures given by the leading English chemist Sir Humphry Davy, he sent Davy a bound copy of the lecture notes he'd taken. The chemist was impressed, and when a position became vacant he offered the young man a job as his laboratory assistant. Davy's gamble on this unknown child of poverty could not have paid off more handsomely.

From the beginning the bookbinder showed an uncanny feeling for both electric and magnetic phenomena. The Danish physicist Hans Christian Oersted had recently discovered that when an electric current flowed through a wire, the wire began to act like a magnet. Faraday reasoned that if electric current could induce magnetism, then the opposite should also be true, and magnetism should induce an electric current. It was an intuitive insight based not so much on reason as on an almost religious faith in the unity of nature. That faith was rewarded when in 1831 Faraday discovered that if he moved a magnet in and out of a coil of wire, a current would be generated in the wire. On the basis of this discovery, he built the world's first dynamo, the forerunner to all electric power generators. Reversing the process, he constructed the first electric motor.

Yet Faraday's momentous discoveries did not immediately launch an electric power industry. For one thing, nobody yet imagined it would be possible to transmit this power over long distances. That realization did not come until the end of the century, when physicists had acquired a formal mathematical understanding of how magnetism and electricity work together. At the core of that understanding lay a brilliant idea of Faraday's. He knew, as does every schoolchild today, that if he sprinkled iron filings on a piece of cardboard placed on top of a magnet, they would assemble into a distinctive pattern. From this, Faraday concluded that a magnet is surrounded by a region of influence, what physicists now call a magnetic field. Similarly, he concluded that electric charges are surrounded by an electric field. Faraday introduced the idea of fields, but he did not have the training to turn his idea into mathematical form. That feat was accomplished by the brilliant Scottish physicist James Clerk Maxwell (1831–1879). Einstein once compared the duo of Faraday and Maxwell with that of Galileo and Newton: In each case the former intuitively perceived ideas that the latter put into rigorous mathematical terms. In each case insight was turned into equations.

Just as the laws of thermodynamics provided a theoretical basis on which to improve the design and efficiency of engines, Maxwell's equations provided the theoretical framework that made it possible to establish a practicable electric power industry. Every time you turn on a light, heater, oven, hair dryer, computer, television, or stereo you can thank Maxwell and his elegant equations. With electrification, modern physics has literally put *power* into the hands of almost every man, woman, and child in the developed world. Electric power revolutionized not only domestic life, but also industry, because instead of each factory generating its own power by burning coal to drive steam engines, power itself could be delivered directly to the factory. It is not possible to describe here the immense industrial transformation wrought by the advent of electricity, but it is fair to say that modern methods of manufacturing are predicated on the availability of power on tap. With electricity, power became a commodity that, like water, could be "piped" into every home, office, and factory, twenty-four hours a day. To us who have grown up with this miracle, it may seem a rather mundane affair, but no other modern technology, except the printing press, has so powerfully transformed daily human experience.

In addition to providing a theoretical foundation for the electric power industry, Maxwell's work gave rise to that other defining modern industry, telecommunications. From his equations Maxwell made the brilliant deduction that *light* was waves of electric and magnetic fields traveling together. Aside from the *electromagnetic waves* that we see, Maxwell realized his equations allowed for the possibility of other waves. Unfortunately, this humble genius died before his insight could be confirmed, but in 1887 Heinrich Hertz generated such waves, which he christened *radio waves*. By the 1920s radio was bringing music, comedy, and drama into the homes of millions (later billions) of people. It was also bringing news. For the first time in history, people could find out about events taking place on the other side of the globe while they were happening. Later, radio waves would also carry pictures, in the form of television.

Steam engines, electric power, and radio were quintessential realizations of Francis Bacon's vision of the operational possibilities of science. By unlocking the secrets of nature, mathematical men were at last learning to command its hidden forces, and to make all sorts of life-transforming technologies. During the late nineteenth century, there was a growing belief that, as Bacon had suggested, our hope for a "better" future lay in science. That optimism was summed up by Alfred Wallace's 1898 book, *The Wonderful Century*, in which this pioneer of the theory of evolution chronicled the successes and failures of nineteenth-century science. Although Wallace acknowledged that science had produced the odd disaster, his general assessment was summed up by his gushing title. Comparing the technological advances of his own age with those of the past, he wrote: "A comparative estimate of the number and importance of these achievements leads to the conclusion that not only is our century superior to any that have gone before it, but that it may be best compared with the whole preceding historical period. It must therefore be held to constitute the beginning of a new era of human progress."

In the early decades of the twentieth century, the idea that science was the key to a "better" future gained increasing momentum. In 1928 the geneticist J. B. S. Haldane wrote that "civilization as we know it is a poor thing. And if it is to be improved there is no hope save in science. . . . Physics and chemistry have made us rich, biology healthy, and the application of scientific thought to ethics by men such as Bentham has done more than a dozen saints to make us good. The process can only continue if science continues." For Haldane, as for Bacon, science was the key both to a better material world and to a more virtuous society. "We are far from perfect," he wrote, but "we do not hang starving children for stealing food, raid the coast of Africa for slaves, or imprison debtors for life. These advances are the direct and indirect consequences of science." According to Haldane, as also Bacon, it was therefore humanity's moral obligation to embrace science.

Yet there was a major difference between the "better" future Bacon had envisioned and the one imagined by nineteenth- and twentieth-century champions of science. Bacon had seen science as ushering in a new age of Christianity—for him it was definitively the servant of religion. But in the late nineteenth century there was a growing feeling that science and religion were incompatible. For the first time in history, champions of science began to construe religion not merely as irrelevant to, and separate from, science, but as its enemy. For these people, science would be our salvation, independently of any Christian framework. For them, science itself was to be the new religion.

The seeds of enmity between science and religion had in fact been planted during the Enlightenment, when men such as David Hume, Denis Diderot, and the marquis de Condorcet had begun to express contempt for Christianity. Nonetheless, historian Ann Braude has noted that "few people viewed science and religion as enemies before the Civil War." Only during the 1860s did hostility become open. The catalyst for this unprecedented rift was the publication in 1859 of Charles Darwin's *The Origin of Species by Means of Natural Selection*. The idea that man was not created in the image of God but had evolved from "lower" life-forms struck deeper at the heart of Christian belief than anything physicists had discovered, and it caused a furor within Christian circles. Under Pope Pius IX, the Catholic Church retreated into theological conservatism, as did many segments of Protestantism, and, in the face of this adverse ecclesiastical reaction, certain champions of science declared all-out war.

The first salvo in this unfortunate saga was fired by John Draper, a professor of chemistry and biology at New York University. After having been personally involved in a bitter dispute with the vehement anti-evolutionist Bishop Samuel Wilberforce, Draper began work on a book entitled *History of the Conflict Between Religion and Science* (1874). From beginning to end, Draper presented the Catholic Church as the enemy of science, "ferociously suppressing by the stake and the sword every attempt at progress." According to Draper, the Vatican was "steeped in [the] blood" of those who had attempted to follow the path of truth and progress. In his litany of the Church's supposed crimes against science, Draper recounted his own version of the stories of Copernicus, Galileo, and Bruno. Copernicus, he wrote, had "incontestably established the heliocentric theory," for which "the inquisition condemned [his book] as heretical." Galileo was "committed to prison, [and] treated with remorseless severity," and Bruno was burnt at the stake for his scientific beliefs. As the reader will now be aware, much of this was a product of Draper's overactive imagination, but readers of the nineteenth century could only assume it was the result of the most scrupulous research—after all, Draper was a respected scientist with the weight of New York University behind him.

Draper's general argument was soon taken up by other scholars, most notably the historian and Cornell University president Andrew Dickson White, who in 1896 published his influential *History of the Warfare of Science with Theology in Christendom*. For White that conflict was nothing less than "a war waged longer, with battles fiercer, with sieges more persistent, with strategy more shrewd than any of the comparatively transient warfare of Caesar or Napoleon." Although White's target was not religion per se but "dogmatic theology," that distinction was easily forgotten, and ultimately his superior scholarship and reasoned tone lent the idea of a war between science and religion tremendous credibility.

White and Draper's aim was specifically to discredit *organized* religion. Draper portrayed the thousand years of Roman Catholic hegemony as an age of intellectual stagnation. A paradigmatic example he gave of the "darkness" of this era was the medievals' supposed insistence that the earth was flat. But as historian Jeffrey Burton Russell has shown, no serious medieval scholars ever believed in a flat earth. Of course there were a few fanatics who did, just as there are today, but the vision of the medievals as flat-earthers is simply a myth. In opposition to the supposed ignorance of the Church and its followers, Draper credited science not only with great advances toward the truth but also for the end of slavery in America and the end of serfdom in Russia. In short, according to him, science had brought dignity and liberty to all those societies that had embraced its light. Draper and White's message was that science was our only hope both for material salvation and also moral salvation. In contrast to the theologian who would keep humanity in ignorance and darkness, White and Draper's noble scientist boldly stood up for truth and progress. Here the ideal was the mythically embodied Galileo—Mathematical Man as the epitome of all that was good in the human spirit.

At the end of Draper's book, he announced that humanity had reached a point where it must choose between Catholicism and science: "The time approaches when men must take their choice between quiescent, immobile faith and ever-advancing Science—faith with its medieval consolations, Science, which is incessantly scattering its material blessings in the pathway of life, elevating the lot of man in this world and unifying the human race." According to Draper, no rapport was possible between the Roman Catholic Church and science, although he did allow that one might be possible with Protestantism. White, however, was scathing toward both churches. The result of such polemics was to create what for many people would seem an irreconcilable tension between science and any form of organized religion. In effect, the West was presented with a Faustian bargain: Give up religion and its promise of an eternal afterlife in Heaven in return for the promise of heaven here on earth. Increasingly the latter choice won out.

Thus, Bacon's vision of a Christian scientific priesthood was transformed into secular terms: Science would "save" us not by helping us create an ideal Christian society but by giving us electricity and engines, cars and planes, radio and television. As philosopher Mary Midgley has noted, in the twentieth century the dream of earthly salvation through science has become immensely powerful—with physicists leading the charge. Time and again they have promised dazzling new technologies: unlimited sources of energy (nuclear fusion, cold fusion, and solar power), supersonic transport, pollution-free battery-powered cars, ever more powerful microchips, space stations, and space travel. In the twentieth century, it has increasingly been to science rather than to religion that we have turned for consolation and for the hope of a "better" future.

The newfound applicability of science in the late nineteenth century transformed it from an almost exclusively gentlemen's pursuit into a professional activity. As it became clear that science could be useful to industry—that indeed there were fortunes to be made by those who could grasp its principles and turn them into products—a great demand for scientifically trained men arose. To cater to this demand, the latter half of the nineteenth century saw the setting up of many of the world's great technical schools: the Massachusetts Institute of Technology, the Royal College of Science in London, and the Technical Institute of Berlin, to name a few. Science was no longer just a noble pursuit but also a powerful career path. Epitomizing this trend was the dramatic rise of the profession of engineering. As people who could effectively apply the principles of science to real-world problems, engineers have been in enormous demand ever since. Today there are more engineers in America than basic scientists. And here once again Mathematical Man has played a central role, because many of the major branches of engineering are, in effect, applied mathematical science: notably mechanical, electrical, civil, and aeronautical engineering. Indeed the mathematical branches of engineering may well be regarded as the Baconian strand of mathematical science.

Yet it was this very professionalization of science that would once again prove detrimental to women's advancement in the field. Although the end of the nineteenth century marked the time when women finally gained access to higher education in science, ultimately the professionalization of the field became yet another barrier to prevent them from participating in the mainstream. In particular, very few scientifically trained women of the nineteenth century were able to work in industry. Just as the male clergy had kept control of the path to religious salvation, male scientists now maintained a firm grip on the path to technological salvation. Nowhere was this more true than in engineering, which to this day remains one of the most male-dominated of all professional fields. If the Pythagorean strand of mathematical science has proven difficult for women to break into, the Baconian strand has certainly proven no easier.

The difficulties facing women trying to break into science in the nineteenth century can be seen in the case of the English physicist Mary Somerville (1780–1872). After receiving only a rudimentary education—just one year at Miss Primrose's boarding school—Somerville (née Fairfax) began to hunger for more. First she taught herself Latin, but later found herself drawn to mathematics. Determined to pursue this mystery, she borrowed her brothers' textbooks and, again, taught herself. Her parents were so concerned at this turn of events that they removed the candles from her room so she couldn't read at night, but Somerville responded by memorizing the books and working on the problems in her head. Only after she had won a medal in a mathematical competition did anyone begin to take her interest seriously.

Somerville's major contribution to physics was to translate Laplace's monumental book on celestial mechanics from French into English—the reverse version of what Châtelet had done with Newton. As had Châtelet, Somerville assisted her countrymen by appending copious explanatory notes, including mathematical derivations. By making this seminal text available in English, Somerville helped to bring Britain back up to date in a field in which it had fallen drastically behind, and for the next century her book was the standard text on the subject for advanced students at Cambridge University. Yet, as a woman, she was not allowed admission to its hallowed halls. In honor of her contributions to science, the Royal Society commissioned a bust of her to stand in their Great Hall. A letter informing her of this decision declared that members would "honor Science, their country and themselves, in paying this proud tribute to the powers of the female mind." Yet their acknowledgment of her mind pointedly did not extend to opening the Society's doors to its bearer.

Finally, however, just after Somerville's death, the situation for women in science at last began to improve. Here, it was America, not Europe, that led the way. After almost two centuries of calls for higher education for women, the first female colleges were opened, and a number of these schools were keen to stress science. One of the earliest was Mount Holyoke Seminary (later College), founded in 1837 by Mary Lyon, a champion of women's education who insisted on science in her curriculum. Before the Civil War, however, most women's colleges were not true academic institutions but ladies' seminaries, where young women trained to be teachers and missionaries. According to historian Margaret Rossiter, "The real impetus towards full collegiate education of women came with the opening of Vassar College" in 1865, and in the 1870s many of the new state universities began to accept women. This move toward coed schools was greatly aided by the secularization of education, as universities passed from clerical to lay control. One of the leaders of this movement was none other than Andrew Dickson White, who to his credit made Cornell an early center for coeducation.

The new women's colleges not only educated women in science, but also employed them to teach. Unfortunately, these early gains did not develop as hoped, because positions at women's colleges remained almost the *only* jobs open to female science graduates in the nineteenth century. Male colleges, such as Harvard and Yale, would not consider having women on their faculties, and neither were the coed colleges willing to take this step. In an age when women were still primarily expected to be mothers and wives, jobs in industry were also virtually out of the question. To quote Rossiter, American society "became far more willing to educate women in science than to employ them." Rather than entering the mainstream with the men, women scientists found themselves relegated to their own separate and distinctly unequal sphere. One consequence of this was that very few women worked at schools, such as MIT, that were at the center of the new technology based education and thinking. Women were thus denied a place in the very enterprise that was now revolutionizing daily life.

The restrictions on women scientists at the time are evident in the case of Harriet Brooks, a physicist at Barnard College. As had been true of male academics in the Middle Ages, female academics at women's colleges in the late nineteenth and early twentieth centuries were not allowed to get married. Although male academics at these schools were now expected to be married, women had to choose between the university and a family. Many women were so happy to have a job that they agreed to these conditions, but in 1906 Brooks challenged this rule. Announcing her engagement, she also declared her intention to continue working. "It is a duty I owe to my profession and to my sex," she wrote, "to show that a woman has the right to the practice of her profession and cannot be condemned to abandon it merely because she marries." By this stage Brooks had published several papers on radioactivity with the leading English physicist Ernest Rutherford, and he and Nobel laureate J. J. Thomson thought she was on the verge of a brilliant career in physics. Yet Barnard's dean, Laura Gill, replied that a married woman was expected to "dignify her homemaking into a profession, and not assume that she could carry on two full professions at once." A man might be able to sustain both a career and a family, but a woman apparently could not. In the end Brooks resigned, after which she did no more physics.

The professionalizing of science also led to the professionalizing of academic science departments. As science began to fragment into dozens of specialized fields, universities began to establish separate departments for each, and soon the science Ph.D. was invented. Thus, graduate schools came into being. Once again, both innovations had the effect of further marginalizing women, because graduate schools were almost always located at the better-equipped, better-funded male colleges. Most women's colleges couldn't afford to offer specialized programs. Yet increasingly, doctorates were becoming necessary for good jobs in science, and without Ph.D.'s women could not even hope to advance.

When the first women applied to graduate schools, they were told bluntly that there was "no precedent" for their admission and were declined. For example, in 1870, Ellen Swallow applied to MIT to do graduate work in chemistry but she was turned down, apparently on the ground that the department did not want its first graduate degree to go to a woman. Similarly, in 1879 Johns Hopkins refused to grant a doctorate to the mathematician Christine Ladd, even though she had sat in on classes and written an excellent thesis. The fight for access to graduate schools raged for over twenty years, until in the 1890s many schools capitulated and opened their doors to women. Yet grad school remained a difficult path for women; by the year 1900 only three women had been granted Ph.D.'s in physics in the United States. Two doctorates had also been awarded to women in astronomy and nine in mathematics. The battle women faced in engineering was even tougher. Although an American woman first received an engineering degree in 1876, when Elizabeth Bragg was awarded one in civil engineering from the University of California, very few women went into the field in the nineteenth century—and that remained the case well into the twentieth century. If science-based technology was going to "save" humankind, women were going to be savees, not savers—just as they had been in Bacon's New Atlantis.

In the final analysis, for all the gains women undoubtedly made during this "wonderful century," equity remained almost as elusive as ever. In 1882 Mary Whitney, an astronomer and pioneer of women's education, wrote: "We are forced to admit that in spite of the wonderful enlightenment of opinion which the last half century has produced in the public mind in reference to women's ability and position, there is still considerable unreadiness to believe that in the higher professions she either can make or will make herself as proficient as a man."

Yet there was one nineteenth-century woman who proved herself in the most spectacular way as proficient as any man: the Polish physicist Marie Sklodowska Curie—the first person ever to win two Nobel Prizes in science. As with Laure Bassi, Curie's achievements are a testimony to what women scientists could have achieved if only more had been given the chance. Curie's contributions were not just theoretical, they also had enormous practical application. Radium, which she discovered, became the basis of a new medical industry, and radioactivity (which she named, and to whose study she devoted her life) would become the basis of new energy and weapons industries. Whether radioactivity has been used for good purposes or bad, no scientific discovery has more powerfully demonstrated Francis Bacon's vision of the scientist as active operator, putting nature's hidden forces at the service of humanity.

From any angle Marie Curie's life (1867–1934) was extraordinary, not just for what she achieved in science but for what she had to go through simply to get into the field. Whereas a poor young man like Michael Faraday had been able to get a position as a laboratory apprentice, no such option was open to a poor young woman. As the daughter of a physics professor at a boys' gymnasium (high school) and the headmistress of a girls' school, Marie Sklodowska grew up in an atmosphere where learning was revered; and in Poland there was a special urgency to keep the flame of knowledge alive, because years of Russian domination had been crushing Polish culture. From her teenage years, Marie belonged to a movement that aimed to use education to save Polish culture until the Russians could be ousted by force. For the sake of her country, she dreamed of going to university to study science or literature. Her sister Bronia dreamed of studying medicine. But in the 1880s Polish universities were closed to women, and the sisters' only hope was to go abroad, preferably to the Sorbonne.

The girls' father, Wladyslaw, had inspired all his children with a love of science, and he would have liked nothing more than to see Marie and Bronia installed at the Sorbonne. But where was the money to come from? Finally Marie developed a plan. She would become a governess and help support Bronia through medical school. Then, once Bronia graduated, she would help Marie with *her* education. Considering the thankless life of a governess, this was an extraordinarily generous offer, and one that Marie fully honored. From 1885 to 1888 she worked far away from friends and family, saving every penny to help Bronia in Paris. These were years of hard work, loneliness, and intellectual isolation. By 1888 Bronia no longer required Marie's support because their father had taken a difficult but well-paid job as the director of a reform school in order to support her himself.

Bronia now urged Marie to save for her own studies, but by this stage Marie was losing heart, and she almost abandoned the dream. In the end she rallied; after two more years of toil she amassed enough funds for a year in Paris. By the time she enrolled at the Sorbonne in 1891, Marie was twenty-four years old—extremely late to begin formal training in science—and she was ill prepared for the rigorous math and physics classes. Thus she had to devote her whole life to study. She also had to watch every penny, because her savings weren't enough to cover her basic expenses. For weeks at a time she lived on bread and tea, and she froze during the winter because she couldn't afford coal. Nonetheless, Marie regarded these years of poverty-stricken studenthood as the most perfect of her life. In 1893 she passed the *licence* in physics (the equivalent of a bachelor's degree today), coming first in her class, and the following year she passed the *licence* in mathematics, this time ranking second.

In that year the greatest love story in the history of science began: Marie Sklodowska met Pierre Curie. At the time they met, both Marie and Pierre considered themselves destined for single lives. After graduation Marie intended to return to Warsaw to look after her aging father and teach science. Pierre meanwhile, at age thirty-four one of France's leading young physicists, was convinced that he would never find a wife who would tolerate his complete devotion to science. He was the first to fall. Almost from the beginning he realized that in this severe Polish girl he had found the woman of his dreams. After a year of pleading, she agreed to be his wife. They would devote their lives to science together.

In 1897 this idyllic dream was thrown into turmoil by the birth of their daughter Irène. As historian Helena Pycior has remarked: "Around the turn of the century, if marriage did not end a woman's career, motherhood almost certainly did." Aside from the fact that Marie did not want to abandon science, in her own words, "my husband would not even think of it." In the end, the solution to their dilemma arrived in the form of Pierre's father, Dr. Eugène Curie, who came to live with the couple and assumed responsibility for Irène, and later also for their second daughter, Eve. By structuring their lives solely around science and family, Pierre and Marie, with the help of his father, both managed to pursue first-rate research careers for the rest of their lives. One cannot help but remark on the magnificent cut of the Curie men. Like Marie's own father, Eugène and Pierre Curie stand out as luminous examples in this or any age.

In the same year Irène was born, Marie became interested in the subject that would occupy her for the rest of her life—radioactivity. Henri Becquerel had discovered this mysterious phenomenon only the year before, and it was a brave gamble for a young researcher to take on such a topic: Although radioactivity had excited interest, few people at the time felt it had much of a future. Becquerel had worked with uranium, but Marie soon discovered that thorium was also radioactive. At this stage she formulated the audacious hypothesis that radioactivity was an atomic property. Her experiments led her to examine pitchblende, which she discovered was four times as radioactive as uranium. Once again the novice research student, without yet a Ph.D. to her credit, made a bold hypothesis—that pitchblende contained an undiscovered element. At the time, chemists believed they had isolated all the basic elements, so to suddenly declare them wrong was a daring suggestion indeed.

At this point Pierre put aside his own research and joined the fray. The idea of a new element excited him as much as it did Marie. For the next four years, working under appalling conditions in a leaking shed, Pierre and Marie Curie painstakingly tracked their quarry through tons of pitchblende. In the end they succeeded in isolating two new elements: polonium, which they named for Marie's homeland, and the immensely powerful radioactive substance they christened radium. The Curies' work opened up new fields of both physics and chemistry, and in recognition of their pioneering efforts, together with Becquerel in 1903 they were awarded the Nobel Prize for physics. It was the greatest possible victory for a woman in science, and both Pierre and Marie were careful to make sure the scientific establishment recognized her role as well as his in the research.

Still, it was Pierre who was offered a new chair in physics at the Sorbonne. Marie was appointed director of the laboratory the university promised to build for him. But on April 19, 1906, just a few months after Pierre took up his position, tragedy struck when he was hit by a horse-drawn cart and killed. At that moment the world lost a truly superior man, and one of the best partnerships in the history of scientific research came to an abrupt end. From now on Marie Curie would have to negotiate the shoals of the scientific world alone.

In spite of her own Ph.D. (awarded in 1904) and the Nobel Prize, Marie Curie was still widely viewed as Pierre's wife. His death presented the scientific community with an unprecedented dilemma: What was to be done with a woman with no husband who now held the most prestigious scientific prize in the world? The French government's first reaction was to assume she would retire and to offer her a widow's pension. But Marie made it clear she wanted to continue her research, and after intense lobbying by friends, including Pierre's brother, she was offered his chair. Thus, in 1906, seven hundred years after the University of Paris opened, it finally had its first female professor.

Marie Curie went on to build the Laboratoire Curie into one of the world's leading research centers. It was here that her daughter Irène (and Irène's husband) discovered artificial radioactivity, for which they received the 1935 Nobel Prize in chemistry. In 1911 Marie herself had also been awarded a Nobel Prize in chemistry for the discovery of radium. No woman in science before or since has been so highly honored. Yet Marie Curie was never accepted into the French Academy of Sciences and throughout her life was dogged by insinuations that the creative work had, after all, been Pierre's alone. In 1920 Harvard's physics department turned down a proposal to grant her an honorary degree: The implication was that the Nobels rightly belonged to Pierre. Although Curie never cared for the spotlight, she bore the brunt of constant, debilitating, and sometimes quite public slander to which no male physicist would have been subjected. As recently as 1971, at a meeting of the American Physical Society, a well-known (male) physicist declared: "If I had been married to Pierre Curie, I would have been Marie Curie!"

The radium Curie discovered quickly became the focus of industrial interest, particularly for its medical applications. But while others got rich from her discoveries, because the Curies had refused to patent the radium extraction process they had so painstakingly developed, she made not a cent from this work. Idealists to the core, she and Pierre gave their knowledge freely to the world. Marie herself was intensely interested in the application of radioactivity to disease and was proud that her discoveries were being used to save lives. True to Francis Bacon's dream of using science for the good of humanity, Marie Curie would definitely count herself as a scientific savior.

Although the most famous mathematical woman of the nineteenth century, Marie Curie was by no means the only one. Foremost among the others were: Sophie Germain, the French mathematician who contributed to the understanding of elasticity; Sophia Kovalevsky, the Russian mathematical genius who won the

Prix Bordin, a kind of nineteenth-century Nobel Prize; Margaret Maltby, the first American woman to gain a European Ph.D. in physics; and Hertha Marks Ayrton, the English electrical engineer, who was the first woman to read a paper before the Royal Society.

As had Curie, all these women had to overcome immense obstacles to participate in the mathematical sciences. But if even a woman of Curie's genius and determination was almost defeated before she set foot inside a university lecture hall, how insurmountable the barriers must have seemed to most aspiring female physicists. What of those countless others? Who knows what talent was squandered because women were not given equal access to education and careers? Who knows what insights and inventions were lost because more women did not participate in the great technological revolution of the nineteenth century? Even today women physicists remain clustered in academe; few work in industry. Furthermore, the engineering wing of applied mathematical science remains overwhelmingly male-dominated. As late as the 1970s women still represented far less than 1 percent of all American engineers, and by 1988 (the last year for which the National Science Foundation has full figures) they were still less than 5 percent of electrical, electronic, aeronautical, civil, nuclear, and mechanical engineers. As we approach the twenty-first century, the Baconian domain of mathematical science continues to bear the separatist imprint of the vision outlined by Bacon in *The New Atlantis*. The fathers of Solomon's House have cast a long shadow indeed.

8

$\neq$ $\oslash$ $\div$ $\leq$ $\leqslant$ ∞ $\doteq$ $+$ $\pm$ $\sim$ $\sqrt{}$ $>$

The Saint Scientific

WHILE MARIE AND PIERRE CURIE WERE COMPLETING THEIR SEARCH for radium, another young physicist was toiling away in obscurity, tracking his own elusive quarry. His goal was not a new element but a new theoretical synthesis. Whereas the Curies' discoveries about radioactivity helped to usher in a new age of thinking about matter, Albert Einstein's equations ushered in a new age of thinking about space and time—and ultimately about the cosmos as a whole. Not since Newton had any physicist illuminated the world around us on such an epic scale. Here at last was the realization of the Pythagorean dream of embracing the heavens in a set of mathematical relations.

Einstein's theory of the cosmos has made him one of the premier icons of our century. The subject of plays, movies, novels, operas, and commercials, he is instantly recognizable: The famous visage can be seen adorning T-shirts, posters, and coffee mugs around the world. Not even Newton was so feted in his lifetime.

Although the seventeenth-century physicist was revered by his fellow Englishmen, it was not until after Newton's death that his ideas were embraced by the world at large. Einstein, by contrast, lived to see the world embrace not only his theories but also himself. A few days after his death a cartoon appeared in *The Washington Post:* a drawing of the earth floating in space, with a sign plastered across it proclaiming in huge letters, "Albert Einstein lived here."

What is it about Einstein that has earned him what can only be called reverence? Why has this wild-haired, droopy-eyed German so captured public imagination? Why, for instance, does every schoolchild know the name of Einstein rather than those of Faraday, Maxwell, Heisenberg, Bohr, or Schrödinger? Why, indeed, has the word *Einstein* come to symbolize genius itself? A significant part of the answer lies in the fact that Albert Einstein put the transcendence back into science. After a century during which physics had become increasingly concerned with the prosaic and the practical, Einstein once again turned physicists' gaze heavenward. With his general theory of relativity, he created an utterly modern and mathematically sophisticated version of the ancient Pythagorean harmony of the spheres. His elegant equations put back onto the scientific agenda the old question of the mathematical form of existence itself. In doing so, Einstein reignited a quasireligious attitude to physics—with respect not to its application but to its content. After a century during which the Baconian spirit had dominated physics, Einstein reintroduced a Pythagorean tone, and thereby precipitated in the minds of mathematical men the reemergence of the idea of a "divine" plan of creation. Although not widespread at first, in recent years this idea has been taken up with a vengeance by Einstein's successors, foremost among them Stephen Hawking. In short, it is to Einstein that we can trace contemporary physicists' obsession with "the mind of God."

Albert Einstein (1879–1955) may not have been a child prodigy, but if there were few external signs of the man he was to become, to the child himself there were early hints of his destiny. When Albert was four or five years old, his father showed him a compass, and sixty years later, in the nearest he ever came to an autobiography, he recalled the "wonder" that it induced in him. "I can still remember," he wrote, "or at least believe I can remember—that this experience made a deep and lasting impression upon me. Something deeply hidden had to be behind things." At twelve, young Albert experienced "a second wonder" when "a little book dealing with Euclidian plane geometry" came into his hands. He later referred to this text as "the holy geometry booklet." The inspiration received from mathematics, which he first encountered in that little book, continued throughout Einstein's life, for it was always in the language of math that he was determined to express his vision of the world. And no one has used its tools to make a more beautiful picture. Although Einstein was by no means a great mathematician, he created a mathematical description of the universe that stands beside the frescoes of Giotto, the portraits of Leonardo, and the Madonnas of Raphael as an aesthetic high point of Western culture.

But before Albert could be an artist, he had to be a student, and school was hell for him. He hated the rigid German gymnasium system so much that at fifteen he dropped out and spent a year roaming about Italy. Yet for all the mythology about him, Einstein was no slouch in his youth. He may not have been an ideal student, but he taught himself math and science sufficiently well that when at sixteen he applied for entry to the prestigious Zurich Polytechnic—still officially two years too young—he passed the math and physics exams with flying colors. Only his insufficient skills in languages prevented him from being accepted. Unfortunately, when he did get in a year later, classes proved almost as tedious as those at school had been, and once again Einstein ignored the curriculum and pursued his own line of study. Thus, like Newton, he was largely self-taught in science. Whereas Newton had found his own way to the works of Galileo, Kepler, and Descartes, Einstein found his own way to the work of Maxwell.

The only hitch in young Albert's admirable display of initiative was that if one wished to be granted a degree at the end of the course, one was required to pass exams. To this end Einstein sought the assistance of his good friend Marcel Grossmann, a fine young mathematician and model student. Grossmann recognized early on the hidden talents of his unorthodox classmate and lent him his meticulous notes to cram from. With this help, Einstein passed his exams.

If Einstein was unimpressed with the teachers at the Polytechnic, they were equally unimpressed with him, and he was unable to get any assistance in securing an academic position after he graduated. The next few years were very difficult as he scraped together a living doing relief teaching at high schools. Yet during this period he returned to physics with renewed passion and soon produced three original research papers. Nonetheless, academe remained unmoved, and still no job offers were forthcoming. The absurdity of this situation was not lost on those few friends with whom he discussed his ideas, and finally, again with the assistance of Grossmann, in 1902 he was offered a job at the Swiss Patent Office in Bern. Now he was free to concentrate on his science without having to worry about where the money for next month's rent would come from—that is, of course, after he had put in a full day's work. Later, when he had become a celebrity, Einstein looked back on this period with fondness and referred to the patent office as "that secular cloister where I hatched my most beautiful ideas." Among them was the special theory of relativity. His position at the time was "technical expert—third class."

While at the patent office, Einstein continued to inform the world about his revolutionary work by publishing papers in the prestigious journal *Annalen der Physik*. The first papers on relativity came and went without a murmur from academe. Not until 1909, after he had completed the special theory and begun working on the general theory of relativity, was Einstein finally offered a position at Zurich University. Today it seems unbelievable that he should have been so shunned by academe. Not because he was "Einstein!" but because by this stage he had produced a dozen original papers, most of which were works of sheer brilliance. Given the low quality of work that all too often propels people up the academic ladder, the reluctance of universities to embrace Einstein in those dazzling early days remains one of the greater absurdities in the history of higher learning. We might wonder why Einstein didn't give up. In fact, during his seven years at the patent office, there were times when he despaired of ever obtaining a university appointment, or even being granted a Ph.D., but one thing he never doubted was the quality of his work. For all the mythology about Einstein's humility, he was a scintillatingly self-confident young man—some have even said supremely arrogant—at least as far as his science was concerned. From the beginning, he knew that what he was doing was of the utmost importance.

Alone at his government desk, Einstein snatched time between his official duties to ponder the same problem that was confounding the world's foremost physicists: the increasingly alarming disjunction between Newton's and Maxwell's physics. Although at first it had seemed that Maxwell's electromagnetic equations would fit harmoniously into the Newtonian world picture, by the end of the nineteenth century it had become clear that the two worldviews were grossly incompatible. Modern Mathematical Man had hit his first major stumbling block.

The dilemma confronting Einstein can be understood by thinking of two cars speeding toward each other on a highway. If one is traveling at 50 miles an hour and the other at 40, then their velocity relative to each other will be 90 miles an hour. In Newtonian physics, as in everyday experience, velocities add together, which is why head-on collisions tend to be fatal. According to Maxwell's equations, the velocity of light in space is 186,000 miles per second. Quite reasonably, physicists assumed that, as with cars, so too velocities would add together with light waves. Thus, if I were traveling at 1,000 miles per second toward a lamp, then the velocity of its light relative to me should be 186,000 plus 1,000, hence 187,000 miles per second. But when two scientists did an experiment to test this assumption, they found that where light was concerned you could not add velocities together. No matter what the motion of the light source or the observer, light always seemed to travel at 186,000 miles per second—relative to everything.

What this implied to the few physicists brave enough to face up to it was that the laws of either Newton or Maxwell would have to be modified. Numbers of the world's top physicists tried to tamper with Maxwell, but so sacrosanct had Newton become that no one was prepared to touch him. In retrospect we can see that several physicists got close to a solution, but in the end they all balked at the extraordinary leap required. As the wheels spun furiously inside academe, down at the patent office, technical expert, third class, Albert Einstein set out to unify the two undisputed masters of modern physics. What drove him was an unshakable faith in the unity and harmony of nature.

The primary question to explain was, how does light *always* travel with the same speed relative to *everything*? If I travel at a *different* speed from you, then how can light travel at the *same* speed relative to both of us? This Alice in Wonderland scenario defied normal logic. Finally Einstein realized the problem lay not with Maxwell but with Newton. Following the great English genius, physicists ever since had insisted on thinking of space and time as absolute. Einstein now saw that if instead of everyone sharing the same space and time, each person occupied his or her own private space and time, then the problem would be resolved, because in each person's *private* space-time the speed of light would remain constant. Thus, he said, time and space are not absolute, universal phenomena; they depend on how fast the observers are moving. The greater the relative velocity between two people, the greater will be the difference in their respective spaces and times. In particular, the faster someone travels relative to you, the more his or her space will appear to contract, and the more his or her time will appear to stretch. In other words, he or she will appear to be squashed and slowed down. This is the special theory of relativity Einstein announced to the world in 1905.

The initial reaction was blank disbelief. How could anyone seriously believe that space and time were private affairs dependent upon velocity? To most physicists this stretched credibility to the breaking point. Yet the problem was that relativity worked. Not only did it explain the puzzle of the constant speed of light, but, using his equations, Einstein was able to make timely predictions that experiments soon revealed to be correct—such as the behavior of electrons in a magnetic field. However disturbing the implications of relativity might be, its sheer empirical success meant that it could not be ignored. The most astonishing prediction of Einstein's theory was the equivalence of matter and energy, the relation signified by the famous equation $E = mc^2$. According to this equation, each speck of matter is a vast reservoir of power, because every subatomic particle can be transformed into a burst of pure energy. At the time Einstein discovered his formula there was no way to test this prediction, but it was demonstrated with appalling thoroughness in 1945. The shadows burned into the walls of Hiroshima and Nagasaki are chilling proof of the equivalence of matter and energy, and hence a powerful symbol of the linkage between space and time.

The irony is that the idea of absolute space and time had been suspect from the start. When Newton introduced the notion, a number of scientists, notably his German rival Gottfried Leibniz, had denounced it. Even Newton acknowledged there were problems with the idea of absolute space. But for him absolute space and time had more than just scientific significance. For Newton, as you may recall, space was God's sensorium, the medium through which the deity was all-seeing and all-knowing. Thus, for him, the absoluteness of space was synonymous with the presence of an absolute God. Absolute space and time provided not only God with an omniscient viewpoint but also physicists. Although they soon abandoned Newton's association of absolute space and time with God, they retained the desire for a godlike framework for reality. The longing of physicists for godlike omniscience was so great that, for two hundred years, most simply ignored the very real criticisms that were leveled at this notion. It took a rank outsider to jolt them out of this pseudoreligious rut. Yet ironically it would be Einstein himself who would restore a godlike perspective to physics.

While still at the patent office, Einstein had become dissatisfied with his early theory. The problem was that although he had managed to harmonize Newton's laws of motion with Maxwell's electromagnetic equations, he had done so only for the special case when velocities are kept constant. The special theory of relativity did not cover variable, or accelerated motion. As early as 1907, Einstein began to dream of extending his theory to the general case of all possible motion. In what must be one of the most brilliant insights in the history of science, he soon realized this would mean incorporating gravity, because a fundamental feature of gravity is that it causes bodies to accelerate. Thus, Einstein's task now was to integrate Newton's law of gravity into his relativity theory. Once again he had become convinced there must be deeper laws of nature waiting to be discovered, and once again he pressed on alone in search of a unifying vision. This time, however, the synthesis he sought was a far more difficult one, and even with his extraordinary ability to see into the heart of physical problems, it took him a decade of nonstop work to find a relativistic theory of gravity, or what is formally known as the general theory of relativity.

Those ten years are proof that physics does not progress by "genius" alone. One is reminded of Kepler's decade-long quest for the elliptic orbit of Mars. Still in the thick of it, Einstein wrote to a friend: "One thing is certain: that in all my life I have never before labored as hard. . . . Compared with this problem, the original theory of relativity is child's play." Rarely has any scientist faced such a formidable mathematical battle, and once again he turned to his friend Marcel Grossmann for help. Grossmann was now an expert on precisely the kind of exotic mathematics that Einstein needed—the tensor calculus—and after teaching it to his old classmate, he joined him in forging a path toward a solution. During the next few years, Einstein would also be assisted by a number of other top mathematicians. Finally, in 1916, he produced a mathematical theory that combined gravity and special relativity. The fruit of his labor was ten compact and beautiful equations—one for each year of effort.

Special relativity had introduced a new conception of space and time but general relativity introduced a new cosmology and ultimately a new perspective on humanity's place in the cosmic scheme. Although it remains one of the most esoteric of physicists' achievements, we have all been deeply affected by Einstein's masterpiece, for this was the theory that put a time line onto existence itself. Indeed, it was general relativity that put the notion of cosmic *creation* onto the scientific agenda.

Since the late eighteenth century, most physicists had thought of the universe as eternal and static—an endless void filled with stars that had been in the same state since time immemorial and would continue in the same state for time everlasting. This worldview was in part motivated by post-Enlightenment physicists' desire to reject the notion of a supernatural beginning. The problem with a beginning was that, however you looked at it, it seemed to necessitate a "first cause" above and beyond nature—a supernatural power, which by definition science could not illuminate. Post-Enlightenment physicists had avoided this issue by simply denying there was a beginning at all. With no beginning there was no need for a Creator, and hence no need to ever cede the field to theologians.

Yet the equations of general relativity specifically suggested a beginning for the universe—a cataclysmic singularity we now call the big bang. At this singularity, it was not just matter that came into being but space and time themselves. Ever since this event, Einstein's equations declared, the universe has been expanding, getting larger and colder. Far from being static and eternal, the cosmos appeared to have a definite history. Thus, general relativity threw the notion of cosmic genesis right into the scientific arena, and Mathematical Man was forced to confront this seminal biblical event on his own turf. But the idea ran so counter to scientific thinking of the day that, for the first time, Einstein lost his nerve. He fudged his equations and added an extraneous term to force the relativistic universe to be static. He later called this "the greatest mistake of my life." Only after the astronomer Edwin Hubble had studied the motions of galaxies and independently discovered that the universe *was* expanding did Einstein return to the original version of his equations. Only then did he accept what his beautiful formulas had been telling him all along.

One of the major tasks for astrophysicists ever since has been to chart the cosmic time line from the big bang (some 15 billion years ago) to today. In doing so, mathematical men have changed the way we see ourselves in the cosmic scheme. In Christian cosmology, humanity had been a part of the universe almost from the beginning, because Adam and Eve were created on the sixth day. But in the time line that emerged from relativistic cosmology, humanity did not appear for fully 15 billion years. For aeons and aeons there was nothing but subatomic particles; later nothing but atoms and stars. Rather than occupying a central place in the cosmos, in relativistic cosmology humanity has been reduced to a tiny blink at the tail end of existence. It is therefore in time, rather than in space, that modern cosmology so suggests humanity's insignificance. Infinite depths of space might make humans and our earth seem puny, but the idea that the universe existed for billions of years without us was a blow to the very heart of Christian self-perception.

In addition to revealing that the universe is finite in time, general relativity revealed that it also has a definite spatial shape. Instead of the formless infinity of the Newtonian world picture, general relativity depicts the cosmos as an elegant four-dimensional form in which space and time are bound together in a synthesis known as spacetime. Before Einstein physicists had seen space as an inert, featureless void, but relativity revealed it as a complex and dynamic structure, bending and flexing in graceful response to the bodies that inhabit it. We can get an idea of this spacetime "landscape" by considering a sheet of highly flexible rubber stretched out like a trampoline. If I put a bowling ball onto the sheet, it will make a depression, stretching the rubber around the place where the ball rests. According to general relativity, this is what a large mass such as the sun does to the "fabric" of spacetime. Every planet, star, and galaxy makes a "depression" in the fabric of spacetime. As the celestial bodies move about, spacetime responds like a rubber sheet to their movements. But unlike the rubber sheet, with its visible depressions, the depressions in spacetime are not visible; rather, they manifest themselves as *gravity*. According to general relativity, gravity is not a quality of matter itself, but a by-product of the shape of spacetime around it—shape imparted to the spacetime by the matter.

The ultimate consequence of general relativity is that the entire universe is shaped by the matter within it. On the universal scale, matter molds spacetime into a specific four-dimensional form. Thus, there is a geometry to the cosmos. Trying to work out precisely what geometry it has is one of the major tasks occupying cosmologists today. It is not the Euclidian kind Einstein encountered in "the holy geometry booklet" (which we learn in school) but a far more exotic type specifically suited to curved surfaces. With his equations, Einstein had achieved what Pythagoras had dreamed of so long ago: He had found a set of mathematical relationships that described the form of the heavens themselves.

From the start, general relativity lent itself easily to religious interpretations. Two and a half thousand years ago, Plato had declared that "God ever geometrizes." No words could argue this view more eloquently than Einstein's equations. And so, despite the profoundly secular climate of early-twentieth-century science, general relativity was quickly encased in a nimbus of quasi-religious discourse. No one contributed to this more than Einstein himself. His linking of God and relativity dates back to at least 1919 when the first test of general relativity was carried out. In that year an eclipse produced ideal conditions to test the theory's prediction that light from a distant star should be bent as it passes by the sun. When Einstein received the cable bringing him the news that his prediction had been confirmed, a student with him at the time asked how he would have felt if it had *not* been. He replied simply, "Then I would have felt sorry for the dear Lord. The theory is correct." The implication was that, with general relativity, he had discovered the plan of creation God *ought* to have used.

Throughout his life Einstein continued to fuel the notion that in his physics he was elucidating the "divine" plan of creation. In 1921 he was told about an experiment whose results appeared to contradict his beloved theory. Again, rather than be concerned, he confidently declared that "the Lord is subtle, but he is not malicious," and again it turned out that the rumored result was not correct. Further, Einstein had very specific ideas about the "Creator's" plan. One of his most famous aphorisms is the oft-quoted "God does not play dice," which pithily summed up his objections to quantum mechanics. But there is another English translation of the same remark (originally uttered in German) that captures more fully the thrust of Einstein's thinking—"God casts the die, not the dice." Here was Einstein's philosophy of nature in a nutshell: The universe is a divine creation, and it is the task of the physicist to discover the mathematical die from which it was cast. It is this essentially Pythagorean dream that inspires Stephen Hawking and others today.

Einstein's linking of God and physics was sincere, for he saw science as a profoundly religious pursuit. He once wrote: "Science can only be created by those who are thoroughly imbued with the aspiration toward truth and understanding. The source of this feeling, however, springs from the sphere of religion." Einstein used the phrase "cosmic religious feeling" to describe his own, distinctly unorthodox, spiritual leanings. Here is what he had to say about it:

> The religious geniuses of all ages have been distinguished by this kind of religious feeling, which knows no dogma and no God conceived in man's image; so that there can be no church whose central teachings are based on it. . . . How can cosmic religious feeling be communicated from one person to another, if it can give rise to no definite notion of God and no theology? In my view, it is the most important function of art and science to awaken this feeling and keep it alive in those who are receptive to it.

As hinted at in this passage, Einstein's cosmic religion was proffered as an alternative to the traditional varieties. Indeed, he was one of those who believed that traditional religions were the enemies of science, and he advocated his "cosmic religious feeling" as the one *true* faith. The foregoing quotation concludes with the sentiment: "A contemporary has said, not unjustly, that in this materialistic age of ours the serious scientific workers are *the only profoundly religious people* [emphasis added]."

In recent years Einstein himself has come to be seen as the embodiment of the scientist as high priest. His cosmological theory, his eminently quotable remarks about God, and his enigmatic statements about the process of science itself have been woven together to create a public persona of the physicist as religious mystic—an image he was the first to encourage. Definingly, he wrote:

The supreme task of the physicist is to arrive at those universal elementary laws from which the cosmos can be built up by pure deduction. There is no logical path to these laws; only intuition, resting on sympathetic understanding, can lead to them. . . . The state of feeling which makes one capable of such achievements is akin to that of the religious worshiper or of one who is in love.

Since Einstein's death a cult of personality has grown up, presenting him as little short of a saint. Countless articles and biographies have presented a picture of a gentle genius, dedicated to world peace, appalled by racism, and committed to truth and freedom. He has also been portrayed as a person so occupied by transcendent thoughts that he abandoned the normal concerns of daily life—a classic sign of a man supposedly dedicated to higher things.

This veritable canonization of Einstein is not just a personal issue; what it represents is the construction of a public face for physics itself. Moreover, the construction of this image has largely been effected by the physics community, for it is physicists who have written many of the most adulatory biographies: Abraham Pais, Banesh Hoffman, and Carl Seelig among them. The perpetuation of this saintly, otherworldly image of Einstein by the physics community represents, I suggest, the way in which many physicists would like themselves to be seen. In an age when many people are hungering for a rapprochement between the spiritual and the scientific, Einstein seemed to achieve an ideal union of the two. As the "saint scientific" he has become the perfect public relations mascot for physics in the contemporary world.

Unfortunately, this picture of Einstein is a distortion. For all his genuine humanism and insight, Albert Einstein was far from being a saint. In the 1980s hitherto unknown evidence of his private life came to light, revealing a man who often behaved appallingly to those who loved him, notably his first wife, Mileva, and their two sons. In his private life Einstein could be arrogant and exceptionally selfish. Furthermore, he possessed a not insignificant streak of misogyny. He wrote to an admirer that "where you females are concerned, your production center is not in the brain."

Elsewhere he declared: "It is conceivable that Nature may have created a sex without brains!" I report these things with a heavy heart, because the architect of general relativity has been my hero since I was a child. Yet the truth is that Einstein was not a saint but an ever-so-fallible man. The continuing portrayal of a near-divine image of Einstein by the physics community is important not for what it tells us about him but for its testimony to an enduring religiosity within that community. The myth of the saint scientific is not simply a literary fiction, but a powerful cultural image that continues to perpetuate a view of physics as a divine or holy pursuit.

Einstein's reignition of a quasi-religious attitude to physics has also had profound consequences for women. Just when women *were* finally breaking into the sciences, this attitude gave renewed vigor to the old view of the mathematical scientist as some sort of high priest. That view has, I suggest, continued to militate against women's advancement in the field throughout the twentieth century, for physics remains the natural science in which women's participation is by far the lowest. According to the American Physical Society, in 1994 women held only 5 percent of academic positions in physics in the United States, and they constituted just 3 percent of full physics professors. Yet in 1990 (the last year for which the National Science Foundation has full figures), women made up 36.9 percent of the United States scientific work force.

Why is it that a century after American women gained access to graduate education, and after a quarter of a century of affirmative action, they have not made greater inroads into this still-so-prestigious science? I believe a significant part of the answer to these questions is to be found in the continuing religious undercurrents of this science: currents that, as we have seen, have been present in mathematically based science from its inception, and that, in the wake of Einstein, have once again grown strong. But before we consider the situation for women physicists today, let us first consider the situation for Einstein's female contemporaries.

It is not coincidental that, in the early twentieth century, the country leading the world in terms of higher education for women,

America, was far from being a world leader in science. This was particularly so in physics. Not until World War II would the United States assume leadership in physical science. Before then the focus remained in Europe (especially in Einstein's Germany), which, for the most part, lagged far behind the United States with respect to higher education for women. Well into the twentieth century, it remained difficult for European women to earn higher degrees in science. Cambridge University, for instance, did not grant a Ph.D. in physics to a woman until 1926, and even then the recipient was not an Englishwoman but an American, Katharine Burr Blodgett. In Germany (the nerve center of European science) most universities did not admit their own women until 1910, even at the undergraduate level, and there were few German academic-preparatory schools for girls until the 1920s. Except for a few rare exceptions, before World War I European women did not have access to advanced scientific training.

As far as physics is concerned, the implications of this inequity have been enormous, because the first few decades of this century were a period during which European physicists almost entirely remade our mathematical world picture. On the cosmological scale, general relativity emerged, and on the atomic scale quantum mechanics was born. This radical new physics painted a picture of the subatomic realm unlike anything physicists had envisaged before. Instead of being made up of neatly defined objects, in the quantum realm particles are waves, and vice versa; and everything operates by laws of chance. (We shall explore this bizarre realm in more detail in the following chapter.) In a few decades, the picture of reality that had been enshrined in science since Newton's day was overturned, and in its place stood a brand new edifice. Yet once again this new world picture was entirely *man*made. Among the founders of quantum mechanics and the early pioneers of relativistic cosmology, there was not a single woman. Given that a large part of physics ever since has been focused on refining the insights of these seminal decades, women's absence from the field during this period was almost as significant as their absence during the seventeenth century.

The resistance to women within the European scientific community in the early twentieth century can be gauged by comparing Einstein's life with that of two female contemporaries—the German mathematician Emmy Noether and the Austrian physicist Lise Meitner. Whatever resistance Einstein himself had faced from the ivory towers of academe pales by comparison with the treatment they encountered.

Like Einstein, Emmy Noether (1882–1935) came from a comfortable German Jewish middle-class family. She followed in the footsteps of her father, a mathematician at the University of Erlangen, and eventually became one of the century's greats. When Einstein was battling with the mathematics of general relativity, she was one of the people recruited to help him. Although primarily known for her revolutionary work in algebra, Noether developed a mathematical idea that has since become central to both particle physics and the current quest for a unified theory of general relativity and quantum mechanics. In personality, as well as family origin, Noether had much in common with her more famous countryman. As biographer Sharon Bertsch McGrayne has described her, Noether "ignored all the feminine conventions of the day. She was overweight, enthusiastic, and opinionated. She was messy, unfashionable, and comfortable. She was also loving, utterly unselfish, and friendly." And she was wedded to her work. But if Noether possessed many of the same qualities for which Einstein is generally lauded, those same traits all too often made her the butt of jokes. Germany's conservative professors did not know how to deal with a woman who so failed to conform to their idea of the feminine.

As did Marie Curie, Emmy Noether lost precious early years because she was a woman. Yet unlike in the Curie household, no one in the Noether family considered that a daughter might warrant any more than the finishing-school education typically given to middle-class German girls, and so, after completing school, she began to take matters into her own hands. As a first step she spent three years in a teacher training program, one of the few educational opportunities available to her. Having completed this course, she turned her sights to the University of Erlangen. At the time Germany did not allow its own women to take degrees, but with the permission of sympathetic professors she was allowed to sit in on classes as an unofficial student, or "auditor." In this Noether was fortunate, for many German professors remained opposed to female students. An 1895 survey had revealed that the majority believed universities were beyond women's mental capacities; as if from the Middle Ages, one academic had declared that "surrendering our universities to the invasion of women . . . is a shameful display of moral weakness."

At Erlangen, Noether had the opportunity to study mathematics, and, like her father and brother before her, she realized this was her calling. Yet not until 1904, after five years as an auditor, could she formally enroll. Hermann Weyl, the famous mathematician and relativist, would later describe her doctoral thesis as "an awe-inspiring piece of work," and Erlangen's examiners were so impressed they awarded her highest honors. But it was one thing for a woman to be educated in Germany; it was another matter for her to be employed. Professional opportunities for women there were even more limited than in America, and for the next eight years after receiving her Ph.D. Noether worked without pay or position under her father's auspices. She supervised doctoral students, gave talks, and continued her own research, which she published in papers now considered classics in their field. As her father's health began to deteriorate, she also took over his duties, still without formal recognition.

In 1914 Noether came to the attention of David Hilbert, who is generally considered one of the greatest mathematicians of all time. When Noether met Hilbert, he and his colleague Felix Klein were engrossed in helping Einstein resolve his mathematical battle for a relativistic theory of gravity. It happened that Noether's area of expertise was one they needed, so they invited her to join their team at Göttingen University. During the next few years, she helped develop elegant mathematical formulations for a number of important concepts in general relativity—still without position or pay. Her only income was a small trust fund established for her by her mother's brothers. The price she paid for her unorthodox choice as a woman was a life of continual financial hardship.

Both Hilbert and Klein saw the injustice of this situation and agitated for Noether to be formally appointed to the faculty. Klein was already a champion of higher education for women, having been a prime force behind Göttingen's decision to award doctorates to several American women in the 1890s. Yet even with the support of these luminaries, opposition to Noether was fierce. During the ensuing debate, most of the professors agreed that a woman's place was in the home, raising good German sons. Finally, however, at Hilbert's urging, the faculty announced it wished to appoint Noether to the junior position of *Privatdozent*. But professors in other departments would not hear of it. As one put it: "Having become a *Privatdozent,* she can then become a professor and a member of the university senate. Is it to be permitted that a woman enter the senate?" Hilbert replied, *"Meine Herren,* I do not see that the sex of the candidate is an argument against her admission. After all, the senate is not a bathing establishment!"

In spite of Hilbert's advocacy, the Ministry of Education agreed with the old guard. Not until 1921, after Germany had lost the war and the political climate had radically changed, was Noether given the position. Still without pay! During her eighteen years at Göttingen, this world-class mathematician was never granted a proper professorship or a proper wage. She was never covered by the civil service system and never received benefits or a pension. Similarly, and again to the disgust of Hilbert, the Royal Göttingen Academy of Science refused to accept her as a member, and she was never credited on the masthead of the international mathematics journal she helped edit.

While at Göttingen, Noether emerged as one of the founders of the discipline of abstract algebra, which in greatly watered-down form would eventually find its way into every school in America as the "New Math." There is no such thing as a Nobel Prize for mathematics, but if there had been, Emmy Noether would certainly have been considered. Mathematicians know Noether for her groundbreaking work in algebra, but to physicists she is known for Noether's theorem. This elegant theorem relates fundamental physical laws of conservation (such as the conservation of energy and momentum) to a mathematical property known as symmetry. It is the insight of this theorem that has now become central to physicists' quest for a unified theory of forces and particles—a single theory that would unite relativity and quantum mechanics into an all-encompassing package.

By the 1930s, Noether was recognized as a major force by mathematicians the world over, but when Hitler came to power in 1933 and began firing non-Aryans from the universities, she was one of the first to go. As a woman, a Jew, a liberal, and a pacifist, Noether was everything the Brownshirts despised, and she soon found herself desperately seeking a post abroad. Unlike Einstein and Hermann Weyl, who had been installed at the Institute for Advanced Study in Princeton, Noether was unable to obtain a research position. In the end she took a post teaching undergraduates at the women's college Bryn Mawr, but it was clear to everyone that she needed a place where she could continue her advanced work. In 1935, just as it seemed the Institute for Advanced Study was on the verge of appointing her, Emmy Noether died as a result of complications from an operation to remove an ovarian cyst. Although most mathematicians do their best work while young, Noether was a slow burner who was still at the peak of what Weyl in his obituary called "the native productive power of her mathematical genius." The world had lost not only a great mathematician but, as Weyl also noted, "a great woman."

Since Lise Meitner's life (1878–1968) was spent directly in physics, in many ways it forms an even starker contrast to Einstein's than Emmy Noether's. Mirroring Einstein's childhood exposure to the compass, Meitner too experienced a childhood revelation. In her case, the catalyst was a puddle of water covered with an oil slick. Catching the light, the oil made the puddle gleam with all the colors of the rainbow. What, she wondered, could be causing this shimmering magic? The answer Lise was given entranced her, and she became convinced that, if she worked hard enough, she too would come to understand nature's laws. Unfortunately, Austria's laws forbade girls from attending the high schools that prepared boys for the university, and so, while Einstein sailed into the Polytechnic, Meitner languished at home.

The only career an Austrian woman was supposed to be interested in was marriage, but after school Meitner failed to show the slightest inclination toward this path. Concerned about her future, her father inquired how she intended to support herself, to which she replied that she wanted to study physics—hardly an answer guaranteed to allay his fears. Nonetheless, he agreed to hire a tutor to prepare her for the university entrance exams. But first, he insisted, she must secure her employment prospects by spending three years gaining a school teaching certificate, just as Noether had done. Meitner later called these her "lost years" and believed she suffered all her professional life from the loss of this precious early time. In 1901, after Austria opened its universities to women, she finally enrolled at Vienna. Though the same age as Einstein, she *started* her university education the year after he *finished* his.

After gaining her doctorate in physics in 1905, Meitner became fascinated by radioactivity, and approached her parents about the possibility of further study abroad. Impressed by her tenacity, her father agreed to give her a small living allowance, and so at twenty-nine (an age when most men in the field would already have established their careers), Meitner left home to begin a life in physics. Her first choice had been to work with Marie Curie, but Curie turned her down, so she turned instead to Max Planck at the University of Berlin. One of Germany's foremost physicists, and the first quantum theorist, Planck agreed to accept Meitner, but it was clear to her on meeting him that he disapproved of women at university. In an 1897 questionnaire he had written: "Generally, it cannot be emphasized enough that nature herself prescribed to the woman her function as mother and housewife, and that the laws of nature cannot be ignored." On meeting Meitner he admonished her: "But you're a doctor already! What more do you want?" To which she modestly replied: "I would like to gain some real understanding of physics." Lise Meitner had no intention of ignoring the laws of nature.

While sitting in on Planck's classes, Meitner teamed up with a young chemist named Otto Hahn to collaborate on experiments in radioactivity. The Hahn-Meitner team wanted to set up shop at Emil Fischer's chemical institute, but Fischer had a policy that no women were to be allowed in his building. In the end a compromise was reached: Meitner could do her the work out of sight in the basement, but she was not allowed upstairs in the real laboratories. Occasionally, the shy Austrian broke this rule and sneaked upstairs, where she hid under the tiers of seats in the amphitheater to listen to a lecture. Together she and Hahn soon became a world-class team, writing nine papers in their first two years of collaboration. Later they would discover protactinium, a rare radioactive compound, and Meitner would discover thorium D.

In 1908 Prussian universities officially opened their doors to women, and in response Emil Fischer opened his building to Meitner. He even installed a separate toilet for her, and eventually became one of her greatest supporters. Thanks to his influence, in 1912 Hahn and Meitner moved to the newly founded Kaiser Wilhelm Institute for Chemistry. Yet Meitner was still working without pay as a guest researcher, a situation that became increasingly untenable when her father died, ending the allowance she received from home. As with Noether and Curie, she too paid the price of severe financial hardship for her determination to pursue a career in science. Only after the University of Prague offered her a job with a proper salary did the Kaiser Wilhelm Society agree to start paying her.

After World War I, the position of women academics in Germany improved greatly. In 1922 Meitner was permitted to give her first lecture at the University of Berlin, and in 1926 she became Germany's first female physics professor. Hertha Sponer had already been an "unofficial" professor, but Meitner was the first official one. By this time she and Hahn had dissolved their partnership, and she continued to work on radioactivity alone. Under her direction, during the 1920s the Kaiser Wilhelm Institute became one of the foremost research centers for physics, rivaling Curie's institute in France and Rutherford's laboratory in England. Einstein called her "our Madame Curie," and during these years she was among those discussed for a Nobel Prize.

Then suddenly, in 1934, the field of radioactivity was thrown wide open, and Meitner found herself at the center of one of the seminal challenges of twentieth-century physics. The Italian physicist Enrico Fermi had discovered that, if he bombarded uranium with neutrons, some of the uranium atoms were transformed into another element. This transmutation of elements was the realization of the age-old goal of the alchemists. The question was, What did uranium get transformed into? Everyone, including Meitner, originally assumed the uranium atoms absorbed the neutrons and became even larger atoms—what are known as transuranic elements. And so the race was on to identify these exotic atoms. Realizing she needed chemical as well as physical expertise, Meitner invited Hahn to collaborate with her again, and together with a young chemist, Fritz Strassmann, they competed with Fermi's team in Italy and Irène Joliot-Curie's team in France to solve the uranium puzzle.

Unfortunately, as a Jew, Meitner was also competing with the rising tide of anti-Semitism. Being a foreigner and working at a private institute, at first she seemed safe from Hitler's purges. But when Germany invaded Austria in 1938, she legally became a German citizen. By having stayed in Germany long after most Jewish scientists had fled, she found herself in a dangerous situation. Fortunately, friends abroad arranged to have her spirited out of the country, leaving Hahn and Strassmann to continue the team's work. Although living in exile in Sweden, Meitner continued to be in almost daily contact with them by mail.

A few months after her departure, Meitner received news from Hahn and Strassmann that they had discovered something very strange. Instead of the expected transuranic elements, they had evidence that uranium atoms had been transformed into much smaller atoms of barium. But how could that be? How could a big atom turn into a small one? They pleaded with Meitner, the physicist, to explain this bizarre finding. A couple of weeks later, Meitner and her nephew, the physicist Otto Frisch, were discussing this enigma while walking in the country when suddenly Meitner had a revelation. What if a uranium atom did not absorb a neutron, but instead was split in two by it? She and Frisch stopped in their tracks and began to calculate on scraps of paper. Yes, it would work. Uranium could be split into barium and krypton. Meitner and Frisch had elucidated the process of *nuclear fission*—a term they coined in a joint paper a few weeks later. It is this process that is the basis of nuclear power plants and the atomic bomb.

The importance of nuclear fission was so obvious that in 1944 Otto Hahn was awarded the Nobel Prize in chemistry. Otto Hahn alone! Neither Meitner nor Strassmann nor Frisch were mentioned. Meitner—who had initiated the team's work, who had been their guiding force, and who had elucidated the theory of nuclear fission—was left out in the cold. The machinations of the Nobel committee have often raised eyebrows, but many physicists agree that Meitner clearly deserved to share this award. Speculations abound as to why she did not, but one cannot but wonder: Had she been a man, would she have been so readily denied this honor? As is that of Marie Curie, Lise Meitner's story is a testimony to the precarious position of women in physics in the early part of this century.

Although the physics community eventually accepted Curie and Meitner, both had to overcome immense resistance to earn their places and respect. Even then they too often had to contend with second-rate treatment. Both the French Academy of Sciences and America's National Academy of Science refused to accept Curie as a member. When Ernest Rutherford came to visit Meitner and Hahn, instead of inviting her to join in the physics talk, he expected her to take his wife shopping. For women scientists in the early twentieth century, marriage also remained a problem. Because few could hope to meet a Pierre Curie, many simply gave up after tying the knot. It is no coincidence that neither Noether nor Meitner married. But a life without husband or children, with few employment prospects, and no job security, was a high price to pay for the privilege of participating in science. If even women of genius could expect little better, it is hardly surprising there were very few women in physics at the time.

Their absence is evident with the quantum physics that emerged during the 1920s and 1930s. Unlike relativity, which emerged from the mind of a single individual, quantum mechanics was the work of a large and internationally dispersed group, among them: Max Planck, Albert Einstein, Niels Bohr, Werner Heisenberg, Erwin Schrödinger, Louis de Broglie, Wolfgang Pauli, Paul Dirac, Max Born, Enrico Fermi, and Satyendra Bose. As in the scientific revolution of the seventeenth century, the picture of reality that emerged from this "new physics" was created by men alone.

Why should it matter whether women were involved in the construction of this world picture? The reason for taking cognizance of this fact is that the quantum world picture is not just a matter of cut-and-dried "science," it has also been the result of human interpretation. I do not mean to imply here that the mathematical relationships the quantum scientists discovered were made up; I am not espousing a purely relativist view of science. But I *am* saying that the way these relationships were *interpreted*—and hence the "reality" that we are told they describe—*was* a cultural construction, just as was the mechanistic world picture of the seventeenth century. This is particularly important in the case of quantum mechanics, because from the beginning there has been fierce opposition within the physics community itself to the standard interpretation of the quantum equations. It is by no means evident even today just what sort of reality quantum mechanics does describe.

Evelyn Fox Keller, the feminist philosopher of science, has argued that women might well have new perspectives to bring to the process of interpreting nature. Keller's point is not that women innately think in different ways than men but rather that because women are often *acculturated* differently they do often have different ways of seeing and interpreting. Given women's different cultural experiences, it is entirely conceivable that they would have something new to bring to the debates about reality that quantum mechanics has generated.

For all the advances women in science had made, by the 1930s very few had penetrated to the interpretive core of the physics community where the mathematical world picture was being remade. And although today there are more women within that core than ever before, their numbers remain small. I suggest that this is not unconnected to the quasi-religious mentality that Einstein reinjected into the discourse about physics. As we shall see in the following chapter, in the latter half of this century the evolution of Mathematical Man's world picture has centered on the effort to unite quantum mechanics with general relativity, and it is in the quest for this all-encompassing union that the religious undercurrent of physics has come bursting to the fore.

9

Ω μ δ Ω ω π Σ Δ γ β α ℍ

Quantum Mechanics and a "Theory of Everything"

ALBERT EINSTEIN HAD NEVER BEEN A MAN OF SMALL AMBITIONS. ALthough he had discovered equations that embraced the cosmos, he dreamed of something even grander, and in the 1920s he began to imagine a theory that would describe not only space and time, but also the matter within it. What Einstein wanted was a single set of equations that would embrace everything in the physical universe: space, time, matter, motion, and force. It was an unbelievably audacious vision, in which every subatomic particle, every atom, every galaxy, every star, every planet, and ultimately every living being would be revealed as a complex vibration in a universal force field. Everything that is would be enfolded into a vast, humming field of energy. Such a theory would be the ultimate realization of Pythagoras' dream, for the entire universe, and everything it contained, would be depicted as math made manifest.

Not immodestly, Einstein imagined he could achieve this synthesis by expanding his general theory of relativity. With that theory he had found equations that encompassed space, time, and gravity, but there were two things general relativity did not include. The most pressing omission was matter itself. Although in general relativity it is matter that shapes spacetime, Einstein's equations could not explain where the matter came from, or what it was. As a solution to this mystery, Einstein imagined each particle of matter not as a hard little mass separate from spacetime, but as a fluctuation, or vibration, in that background field. The second omission from general relativity was an explanation for the electromagnetic force. Just as Einstein had previously shown that gravity is a by-product of the shape of spacetime, so he hoped to show the same for electromagnetism. In Einstein's vision, both these forces and matter would all be explained as manifestations of spacetime, by-products of the all-encompassing underlying fabric of existence. Einstein called his vision a "unified field theory" and spent the last four decades of his life searching for such a synthesis. But in this he failed; for once his legendary ability to see into the heart of nature could not guide him to his longed-for destination.

The problem was that Einstein refused to believe that a unified theory of physics would need to encompass the insights of the emerging field of quantum mechanics. He believed he could come to a grand synthesis of force and matter simply by extending the scheme of general relativity and ignoring the casinolike machinations of the quantum realm. Because in Einstein's view "God does not play dice," for him an ultimate theory of the universe could not possibly include this hated aspect of the quantum mechanical world picture. Yet as time wore on it became increasingly clear that God was indeed a gambler.

No theory in the history of science has been more empirically successful than quantum mechanics. On the strength of quantum mechanics, humanity has built the microchip industry, and hence the computer industry. An understanding of the quantum realm has also given us the laser, and hence fiber-optic communications, CD players, bar code readers, laser surgery, laser-guided weapons, and in the future probably also optical computing. Quantum dots for building designer atoms, quantum interference devices for measuring brain function, and quantum cryptography are some of the technologies in the pipeline. Whatever Einstein's philosophical objections to the quantum world picture, by the end of the 1930s it was clear that no unified theory of physics could ignore the lessons from this domain.

Einstein's objection to quantum mechanics was that it portrayed a picture of reality utterly at odds with general relativity. Whereas the relativistic world picture was smooth and continuous, the quantum world picture was jagged and disjointed. In the relativistic world, objects move smoothly—a ball rolls across a field, and a planet revolves around a sun—but, in the quantum world, subatomic particles lurch about, suddenly disappearing from their starting points and reappearing as if by magic somewhere else. In the subatomic realm, everything happens by chance. The mathematical equations that describe the quantum realm do not allow physicists to make firm calculations about the future, as they can do with the motions of the celestial bodies. Rather, quantum equations allow one to calculate only the *statistical odds* of various possible outcomes. Indeed, dealing with the quantum realm is like playing a game of craps in some bizarre casino.

In the quantum realm, particles are not nice neat objects with well-defined trajectories; instead, they are wavelike entities with hazy and ill-defined paths. In many cases you cannot watch a subatomic particle move from A to B; you can only observe it at point A, and, sometime later, observe it again, at point B. Just how it got there is a mystery. In this realm, particles sometimes act entirely like waves, and vice versa. This equivalence of particles and waves is related to the equivalence of matter and energy that Einstein discovered. Furthermore, in the subatomic domain, all things come in discrete packets, called quanta. The packets of matter are the various types of particles—electrons, protons, quarks, and so on; the packets of energy are called photons. It is these quanta that behave according to rules of probability. Where general relativity depicted the cosmos as a stately manifestation of graceful geometry, an elegant mathematical waltz, quantum mechanics depicted the subatomic realm as a strange stochastic jitterbug. How could nature be both things at once? How could both pictures be right? Yet how could either be wrong?

With the advent of general relativity and quantum mechanics, Mathematical Man's world picture was cleaved in two. It was as if there were two entirely separate realities coexisting in the one space. The unity of the Newtonian world picture had been replaced by a bipolarity. Yet surprisingly, from the 1920s to the 1960s most mathematical men were too drunk on the success of quantum mechanics to care that its vision of reality clashed with that of relativistic cosmology. But if most quantum physicists were untroubled by the schism at the heart of their science, Einstein was tormented by it, and the second half of his life is littered with the corpses of failed attempts at a unified theory. During these years many physicists came to view his efforts as the crackpot obsession of a genius who'd gone off the rails.

It must have taken immense faith for Einstein to go on believing, in the face of so little success, that a unified theory was possible. We may now appreciate the full force of his remark that "the search for harmony is the source of the inexhaustible patience and perseverance with which [the physicist] devote[s] himself." In this case, perseverance did not pay off, yet Einstein never stopped looking for the cherished synthesis. Today his quest for a unified theory has become a veritable obsession with many mathematical men. Whereas Einstein evoked amusement for his stubborn adherence to this dream, that same dream now occupies the energies of a good percentage of the world's top physicists. It has even been given a name: a theory of everything, often simply referred to as a TOE. As its grandiose, and rather Monty Pythonesque name suggests, a theory of everything is contemporary Mathematical Man's attempt to wrap up existence into a tidy logical package. Nobel laureate Leon Lederman has voiced the hope that physicists will eventually be able to express this theory as a single equation you could write on the back of a T-shirt—a sartorial adornment that would fulfill the essentially Pythagorean quest for a universal mathematical "harmony."

In his quest for unification, Einstein wrestled with two forces, gravity and electromagnetism, but today's TOE seekers must unify *four* separate forces, because in the 1930s physicists discovered two more forces operating inside the nuclei of atoms—the weak nuclear force and the strong nuclear force, which will be discussed in more depth shortly. These forces had not been detected earlier because they only function at the scale of the atomic nucleus. Along with electromagnetism they are also quantum forces. Physicists now believe these four constitute the full complement of nature's "fundamental" forces. Among them they encompass the gamut of physical phenomena: the cosmological (gravity), the atomic (electromagnetism), and the nuclear (the weak and strong forces).

Yet although these forces are each very different, TOE physicists believe that ultimately they are various manifestations of one all-powerful force, often designated the superforce. A theory of everything would be a mathematical account of this superforce. This integration of the lesser forces would have the power and potency of them all. As English physicist Paul Davies has put it, the superforce is "the fountainhead of all existence." Protons, pulsars, and people, we are all supposedly by-products of this all-embracing power, and a theory of everything would be the mathematical relationship that describes its operation.

While the idea of one all-embracing force governed by one all-embracing law of nature seems such a contemporary vision, it was not in fact a new dream for Mathematical Man. Two hundred years before Einstein, a Jesuit priest named Roger Boscovitch had a similar vision, and, in the middle of the eighteenth century, he suggested how such a force might work. His work was the first attempt at a mathematical theory of everything. Developed a century before Maxwell's elucidation of the electromagnetic force, and fully two centuries before the discovery of the nuclear forces, the particulars of Boscovitch's theory could not have stood the test of time, yet in spirit and scope it prefigures current efforts. Indeed, there is a direct chain linking him to today's TOE champions, a chain whose links include many of the greatest figures in the history of modern physics, starting with Newton. And as with Newton, Boscovitch was a deeply religious man whose physics resonated with powerful theological undertones. Whereas contemporary physicists who talk about the "mind of God" are following in Einstein's footsteps, he in turn was following in Boscovitch's.

Like Copernicus, Roger Boscovitch (1711–1787) was a Slav.

Today he is claimed with equal vigor by the Serbs, the Croats, and the Dalmatians. His nationality is surely a significant part of the reason that this visionary physicist isn't more famous today—it is difficult to imagine that any Anglo scientist of such caliber would have remained so outside the spotlight. Most of Boscovitch's work has still not been translated from the Latin, and many of his papers are still awaiting scholarly assessment. Most contemporary histories of the science do not mention him at all, yet from the time he published his major work, the *Theory of Natural Philosophy*, in 1758 until the late nineteenth century, Boscovitch's ideas exercised an immense influence. Ironically, it has been only in the present century, when his ideas have finally come into their own, that he has dropped out of scientific consciousness.

Boscovitch was born Rudjer Josip Bošković, in the Republic of Dubrovnik, but it is by the Anglicized version of his name that he is more commonly known. The youngest boy among nine children, he grew up in a clever, cultivated, and comfortably well-off family. There were poets on both sides of his lineage, and Boscovitch grew to be a fine composer of verse, penning a literary homage to science that has been described as a fusion of Newton and Virgil. But for all their cultural leanings, the Boscovitches were a pious people, devoutly Roman Catholic like many of the inhabitants of their small prosperous republic, and by the age of fourteen Roger had determined upon a career in the Church. He made the decision to enter the long and difficult program of intellectual and religious training that led to ordination in the Jesuit Order. His brother Baro also became a Jesuit, brother Ivan a Dominican, and sister Marija a nun in the Dubrovnik Convent of St. Catherine.

After a rigorous two-year period of spiritual testing, the young novitiate took vows of poverty, chastity, and obedience and began his formal training, which with luck would end in priesthood some fifteen years hence. Boscovitch entered the Jesuit fold in 1727, the year Newton died. Little could anyone have known that this intense Slavic boy would pick up the master's melody and craft it into his own mathematical symphony. Although in the sixteenth century the Jesuits had been leaders in the new scientific movement, by the eighteenth century the order had fallen well behind. With his commitment to Newtonian physics, Boscovitch would strive to bring them back to the front rank of science, and would eventually become one of their most illustrious scientific representatives.

From an early age Boscovitch exhibited brilliance in both mathematics and science, but unlike Newton, he was not free to devote himself to these interests. From the moment he was ordained, he was bound by the Jesuit code to serve the Society as its leaders decreed, and they decided his talents were needed in more political arenas, for Boscovitch was also a skillful diplomat. As the anticlerical climate of the eighteenth century intensified, the Jesuits had much need of diplomatic acumen, and Boscovitch was often called to represent the order at various European courts. He spent much of his life traveling around Europe on Machiavellian political missions. Rather than begrudge these distractions, Boscovitch reveled in his cosmopolitan role. A brilliant and witty conversationalist, a raconteur and poet who could extemporize at length, he shone on the social scene and thoroughly enjoyed mixing with the secular elite.

Yet Boscovitch still managed to pursue his scientific work, turning out a vast range of papers and ideas. Given his diplomatic obligations and hectic social life, it is incredible that he had time to do so much. Being a priest, he did not have the demands of a family, but few people can claim such a wide field of achievement. Boscovitch did original work in astronomy, geodesy, mathematics, optics, and the theory of matter. He had a sound knowledge of engineering and was called upon to offer advice on the preservation of several major churches, including St. Peter's, whose dome was feared to be in danger of collapse. He devised a scheme to drain the pontine marshes, designed an astronomical observatory, improved telescope lenses, made measurements to determine the size and shape of the earth, and was an expert on mapmaking. During his life he published over a hundred books and papers. But above all Roger Boscovitch dreamed about a universal force of nature.

In the early eighteenth century, the only force that had been identified by physicists was gravity. But it soon became clear that gravity could not account for the behavior of things in the atomic domain. At the time the idea of atoms was still highly contentious, but many scientists were convinced that the diverse properties of materials must be caused by various arrangements of different types of atoms. The problem was that no one could successfully explain what caused atomic arrangements. What force was responsible for order in the atomic domain? Gravity alone did not seem to provide a satisfactory answer.

One of the shortcomings of gravity was that, left to its own devices, it tends to pull things into basically spherical clumps. Stars and planets, for example, are all more or less spherical. Yet people believed that in order to account for the wide range of materials, atoms had to form clusters (what we would now call molecules) with many different shapes. The shape of atomic clusters supposedly determined the multitude of a material's properties: hardness, density, color, and so on. Boscovitch reasoned that, in order to achieve a range of molecular shapes, there must be a repulsive force acting in concert with the attractive force of gravity; molecular form would then arise from the balancing of attractive and repulsive forces between individual atoms. Yet on contemplating the immense diversity of material properties, Boscovitch decided that two forces were insufficient and that a number of attractive and repulsive forces would be necessary. But, although he was proposing a multitude of atomic forces, Boscovitch firmly believed in the unity of nature, so he suggested that all these atomic forces, along with gravity, must be aspects of one all-encompassing universal force. Furthermore, this superforce must be described by a single universal law of nature.

The essence of Boscovitch's force was sublimely simple: It was to be a generalized version of Newtonian gravity. The Slavic priest was an immense admirer of the English physicist, and what greater homage to Newton could there be than to extend his gravitational law into the atomic domain? To see what Boscovitch had in mind, imagine two atoms floating in space. According to Newton's law, the force between them is always attractive, but Boscovitch modified this by proposing that when the atoms get very close to each other, the attraction turns into a repulsion. Closer still, it again becomes attractive. As the atoms get closer and closer, the force between them oscillates between attraction and repulsion. Thus, according to Boscovitch, one single oscillatory force is responsible for both cosmological and atomic action. Today physicists know that a unified force of nature must act in a far more complex manner; nonetheless, Boscovitch's vision of a universal force, embracing both gravity and a variety of atomic forces, lives on in the superforce.

In the following century Boscovitch's ideas exerted a powerful influence on Michael Faraday. Although today Faraday is remembered largely as an experimental physicist, he viewed himself as a "natural philosopher." Behind his discovery of electromagnetic induction, and his subsequent invention of the electric motor and dynamo lay a deeper quest. Faraday believed wholeheartedly in the unity of nature, and following Boscovitch's lead, he believed that all the forces of nature must be expressions of a single universal force. Faraday's whole scientific career can indeed be seen as a search for unity among the forces.

By Faraday's time, physicists had recognized three forces: gravity, the electric force, and the magnetic force. Many scientists believed that light was caused by yet another force. Convinced that they must all be related, Faraday searched endlessly for links between them, and no one has done more to advance the program of unity. Not only did Faraday show that the electric and magnetic forces were bound together, he was the first to suggest that they were also the basis of light. Faraday also demonstrated, through his seminal experiments in electrolysis, that the force holding matter together was none other than the electric force. Today we know that it is the positive and negative electric charges within atoms that hold them together and bind them into molecules. Just as Boscovitch had suggested, the forms of the atomic realm *are* caused by complex balances between attractive and repulsive forces.

Faraday's work revealed that electricity, magnetism, light, and matter are all inextricably linked by invisible influences. Furthermore, he laid the groundwork for a formal understanding of these influences with his notion of what would later be called "force fields." But few physicists in the nineteenth century were interested in the idea of a unified force of nature, so, although Faraday was respected as an experimentalist, the physics community ignored his theoretical speculations about unification, tending to regard them as crackpot fabulations. A century later Einstein would feel the same cold wind from his colleagues. Faraday's belief in the unity of the electric and magnetic forces and light was of course vindicated later in the century by James Clerk Maxwell, who showed that they were all just aspects of one encompassing *electromagnetic* force.

From the eighteenth to the twentieth century, the dream of a unified theory of the forces of nature was passed like a baton in a relay from Boscovitch to Faraday to Maxwell to Einstein. These four physicists, all of whom departed from the orthodox thinking of their day, had transformed physics and given it a *goal*—one that has, in the last three decades, finally been endorsed by the physics community at large.

It is far from insignificant, I suggest, that all these early champions of a unified force of nature were deeply religious men, particularly the first two. Boscovitch was a priest, and Faraday a member of a tiny but intense Christian sect, the Sandemanians. So serious were the Sandemanians that on one rare occasion when Faraday missed a Sunday service—because he had the even rarer honor of being invited to lunch with Queen Victoria—the sect's council demanded that he repent. Proving he too could be stubborn, Faraday refused, and as punishment he was ostracized for a time. Nonetheless, he remained a lifelong devotee and served on the Sandemanian rotating council of elders. Maxwell was also a devout believer, who regarded his Christian faith as a vital part of his life, and Einstein, as we have seen, was deeply committed to his own idiosyncratic religiosity. The idea that there *must* be one force ultimately responsible for all action and form in the universe can be considered as a scientific parallel of monotheism. I do not think we should regard it as a mere coincidence that such a goal has arisen within the Judeo-Christian culture. The longing for one all-encompassing cosmic law is, I suggest, the scientific legacy of more than three millennia of faith in one all-encompassing principle known as God. The very fact that this idea was introduced into modern physics by a priest is not to be ignored.

That Einstein did not succeed in unifying gravity and electromagnetism now looks much less like a personal failure than it must have seemed in his lifetime, because no one has yet found a way to unite these two forces. Furthermore, since physicists have discovered the two nuclear forces, it is now known that the entire problem of unification is a great deal more complex than Einstein imagined. Ironically, although gravity was the first force to be discovered, it will probably be the last to be brought into the unified fold. What neither Einstein nor Boscovitch could possibly have known was that rather than being a good model for understanding the forces of the atomic domain, gravity is the odd man out. History had dealt them both a perverse hand. In the late twentieth century, most of those who have been working toward unification have concentrated not on gravity but rather on the three "atomic" forces—electromagnetism, the weak nuclear force, and the strong nuclear force.

Of the two nuclear forces, the strong force was the first to impinge upon physicists' consciousness. By the early 1930s, experiments had revealed that the atomic nucleus was not a hard, solid lump, as originally believed, but a conglomeration of separate particles—some positive (the protons) and some neutral (the neutrons). The discovery of the conglomerate nature of the nucleus raised a serious problem, because the natural tendency of any collection of particles with the same charge is to repel one another. As with magnets, so with electric charges: Opposites attract, but likes repel. Why, then, didn't nuclei blow themselves apart? What was holding them together? In 1935 the Japanese physicist Hideki Yukawa proposed a solution. He suggested that inside the nucleus there is another powerful force that overcomes the electric repulsion between the protons and binds them together. Yukawa's force came to be known as the strong nuclear force. Not a particularly poetic appellation, but certainly to the point. (Later, the strong force would be associated with the particles known as quarks, which physicists have since discovered are the basic constituents of both protons and neutrons.)

With the elucidation of the strong force, the set of known forces formed a pleasing pattern: Gravity ruled the cosmological domain, the electromagnetic force ruled the atomic domain, and the strong force ruled the nuclear domain. Yet this neat arrangement turned out not to be the full story, because certain particles, including the neutron, were found to be unstable. It is the instability of neutrons that gives rise to radioactivity in elements such as uranium. Physicists soon realized the strong force could not account for this instability, so they proposed a second nuclear force, which by comparison with its sibling they dubbed the weak force. Fortunately, two nuclear forces seem to be sufficient, and today most physicists feel confident that we live in a universe in which four, and only four, forces account for the full range of nature's activities. Given Einstein's difficulty in uniting just two of them, this hardly seems cause for jubilation, but let us not forget that the situation might have been worse. There might have been seven forces, or seventeen, or seventy. Four is a not too daunting number, and one is reminded of Einstein's remark that "the Lord is subtle but he is not malicious." The existential puzzle posed by the four forces is far from trivial, but at least there are grounds for hoping its resolution is not beyond the bounds of human intellectual capability.

The puzzle, however, is rather more daunting than an initial perusal might suggest. Not only must a unified theory unite the four forces but, because forces and particles are intimately intertwined, any such theory must also explain the diversity of particles. In this case "the Lord" may well be malicious, for, since physicists have been digging into the bowels of matter with their accelerators, they have discovered a positively alarming number of these: several hundred in fact. The wild proliferation of subatomic particles during the 1950s and 1960s prompted Enrico Fermi to remark that if he could remember the names of them all he would have been a botanist.

The intimate pas de deux between forces and particles arises out of the fact that forces are the means by which particles interact, and particles are the entities that forces act upon. Furthermore, in contemporary quantum theory, forces themselves are described in terms of force-carrying particles known as bosons. Photons, for example, are the force-carrying particles (or bosons) of the electromagnetic force. Because TOE physicists are wedded to the belief that at its core nature is simple, they find the notion of several hundred "fundamental" particles inherently distasteful. As Paul Davies has noted, "the very attractiveness" of the idea of fundamental particles is that they "need come only in a handful of varieties. The complexity of matter is then explained as arising not from the multiplicity of ingredients but from the multiplicity of combinations." Physicists thus believe there must be some less numerous but "more fundamental" set of particles than current evidence suggests. Because particles and forces are two sides of the same coin, the quest for "truly fundamental" particles has therefore become part and parcel of the quest to understand the linkages among the forces themselves.

Fortunately, today's physicists can do what Boscovitch could only dream about: study "atomic" forces directly. The tools they use to do so are particle accelerators, and with the help of these machines they have at last begun to make progress toward a unified theory. In the 1960s, a better understanding of particle interactions led physicists to develop a theory that unified the electromagnetic and the weak nuclear forces. Just as Maxwell had shown in the 1860s how the electric and magnetic forces could be seen as different aspects of one encompassing electromagnetic force, so Steven Weinberg, Abdus Salam, and Sheldon Glashow showed that the electromagnetic and weak nuclear forces could be seen as different aspects of a more encompassing electroweak force. As the Age of Aquarius dawned, the number of forces was reduced from four to three.

The success of electroweak unification spurred physicists to start thinking about how they could unite this combined force with the strong nuclear force. Theories that attempt to do this are called grand unified theories, generally referred to as GUTs. It is an unfortunate name for what is really a rather beautiful idea: a single law, describing a single force, from which emanates the entire quantum realm. This would, in effect, be a theory of everything for the subatomic domain. At the moment there are many potential candidates for a GUT, all involving horrendously complex mathematics, and physicists are now trying to work out which, if any, might be viable.

But even as theorists struggle with GUTs, some are pushing the math further to attempt the unification of all four forces. It is these grand-slam efforts that are called theories of everything. In a successful TOE, not only all forces, but also all particles would be united—all would be one and one would be all. So, not only is the superforce supposed to be the supreme power of the universe, according to its champions, it is also the supreme substance—the distillation of all the different particles, the very quintessence of matter. Thus, a TOE would meld force and matter into one supreme law. The prime candidates for such a theory go by the name of superstring theories. As with GUTs, these are all horrendously complex, and the task of working through their labyrinthian equations is currently occupying some of the best mathematical minds of our time.

Whatever its final form, according to its champions, a successful theory of everything would be the ultimate mathematical description of reality. Steven Weinberg has called it "the final theory" of nature, and, he says, such a theory would "bring to an end . . . the ancient search for those principles that cannot be explained in terms of deeper principles." Encompassed in its embrace would be not only all the forces and particles, but space and time as well. As is gravity in general relativity, in a theory of everything, all four forces would be revealed as manifestations of the underlying geometry of spacetime. Similarly, all the particles would be revealed as microscopic vibrations in this field. Just as Einstein imagined, all would be explained as by-products of the universal fabric of spacetime—now also cast in a quantum mold.

The task of finding a mathematical framework to encompass both the quantum and the relativistic perspectives has been truly daunting, but TOE physicists now believe they have found the solution in a conception of the universe that contains not the usual four dimensions (three of space and one of time), but ten. That's how many superstring theory demands. According to current thinking, we do not normally perceive the extra six dimensions because instead of being extended on a cosmic scale, like the regular three dimensions of space, they are very tiny. We may understand this rather startling proposition by considering a long and very thin hose.

From a distance a thin hose would look like a line, but as you got closer you would begin to see that it also had thickness. There is in fact a *second* dimension to the line, in which each point on it is really a tiny circle. By extending this notion to higher dimensions, one can imagine that each point in regular three-dimensional space is not in fact a point but a multidimensional version of the tiny circle. Such a construct is known as a hypersphere. According to superstring theory, each point in regular space is a tiny six-dimensional hypersphere, and it is the properties of this hypersphere that generate the various quantum forces and subatomic particles. In a nutshell, then, Mathematical Man's current world picture depicts a ten-dimensional universe consisting of four "extant" dimensions and six "compactified" ones. The four extant dimensions determine the form of cosmos, and the six compactified dimensions determine the forms of the subatomic domain.

It is an extremely elegant picture, and it's not hard to see why physicists become enamored of it. The question is, Does this lovely mathematical construct bear any relation to the real physical world? It is all very well for TOE physicists to talk about ten dimensions, but the fact remains that nobody has yet detected anything beyond the regular four. The burden of proof rests heavily on the TOE community. The major problem they face is that the effects of unification become apparent only at extremely high temperatures —so high they make the interior of the sun look like the Antarctic. That is why we observe four seemingly different forces rather than one unified superforce. Temperatures in the universe today are too low for unification. Physicists who want to see that unity and verify their ten-dimensional theories have to raise the temperature. They have to generate enough heat to "melt" the separate forces together. In this sense accelerators can be seen as giant heaters: They are used to smash particles together at such high energies that during collisions enormous temperatures are generated—thereby creating the conditions for unification.

The great triumph of this approach occurred in 1983–84, when physicists at the European Center for Nuclear Research (CERN) detected the W and Z bosons predicted by the electroweak theory. Mathematical men now had confirmation of the first stage of unification. But the next stage, the GUT stage, requires such high temperatures there is no way with our current technology that humanity could build a sufficiently powerful accelerator. Although one might imagine that such energy could become available to us with some far-in-the-future technology, full TOE unification would require almost incomprehensible amounts of energy. Fortunately, nature is not bound by our limitations, and it has provided the ultimate particle accelerator, one in which total unification was achieved—the big bang.

According to physicists today, all the matter in our universe was created by this seminal event. The big bang was literally the greatest particle experiment ever. The enormous energy out of which this matter came was initially compressed into a tiny spark of heat so intense that total unification reigned. So even if physicists can't study unification in their own accelerators, they believe they can study it by looking backward in time to nature's accelerator. The further into the past you go, the smaller and hotter the universe gets. Thus, TOE physicists believe that as one approaches the big bang, the ambient temperature becomes hot enough first for electroweak unification, then for grand unification, and finally, total unification. As one gets closer to the big bang, the forces coalesce into the superforce.

Looking at this process going *forward* in time, we see how physicists now believe our universe came into being. The quest for a unified theory of the four forces has thus given rise to the modern scientific account of genesis. "In the Beginning," the story goes, the universe was created in a state of perfect unity. Nothing existed except the unbroken wholeness of the superforce. There was no matter, no particles, no gravity, no electromagnetism, and no nuclear forces. But this perfection reigned for a mere split second, because the violent explosion of Creation caused the budding universe to expand, and as it expanded it cooled, shattering the original unity. First the superforce split into two as gravity separated from the quantum forces, then the strong force separated from its quantum siblings, and finally the weak and electromagnetic forces broke apart. As the forces separated, the various particles came into being. Energy coalesced into matter. When the universe cooled even further, these particles joined to form simple atoms of hydrogen and helium, which slowly congregated into vast clouds. Over the aeons these clouds condensed into galaxies, stars, and planets. Thus, according to modern Mathematical Man, our universe came into being.

In technical terms, TOE physicists speak about this fragmentation of the superforce as a process of "symmetry breaking." In the beginning, they say, there was "perfect symmetry," but it soon shattered, and ever since the universe has been a place of "broken symmetry." This discourse of symmetry refers to mathematical properties of the theory itself—it is this quality that Emmy Noether dealt with in her famous theorem. In physical terms, to say the universe began in a state of perfect symmetry is to assert that in the beginning there was no differentiation. At the quintessential moment of unity, all four forces and all ten dimensions were equal. Like a perfect sphere, the universe would have looked the same in all "directions." Symmetry is essentially a mathematical formalization of the property of sameness. Those who seek the superforce believe that the original spark of creation was a state of undifferentiated sameness. What they hope to show is that the myriad multiplicity in the universe today is not a true multiplicity but the shattered shards of an original "perfect" unity.

It does not take a great leap of imagination to recognize a distinctly Christian undercurrent in this scenario. Like the expulsion of Adam and Eve from the biblical Garden of Eden, TOE physicists have envisioned the history of the universe as a decline from a state of original perfection. And as in the Christian tradition, they yearn to return to this seminal state of "grace." This is their mathematical Eden. Again, I do not think we should view this undercurrent in the TOE quest as a mere coincidence. Physics does not happen in an asocial vacuum, and physicists' dreams are also informed by long cultural traditions. Why, we must ask, have they become so entranced with this particular dream? Why have they come to believe with such faith, that the universe *must* have begun in a state of undifferentiated sameness, and that everything *must* ultimately have been caused by a single all-powerful force?

Such a dream is by no means obvious, for even with high-energy accelerators, physics continues to reveal not a unity but a bewildering richness of multiplicity, including the ever-expanding plethora of subatomic particles. In the face of such manifest evidence to the contrary, it requires an enormous leap of faith to believe that underneath this abundance there is a single principle. I suggest that this faith cannot be justified on purely scientific grounds but must also be seen as in part arising from cultural traditions, in particular the Judeo-Christian tradition of monotheism. After all, there is no a priori reason why the universe might not have been formed according to a multiplicity of principles—a scientific version of polytheism. That is not to say there is no scientific evidence that points to a unity among the forces; there is certainly some, but it is far from overwhelming, and TOE physicists' almost fanatical faith in unity transcends the evidence currently available. The very fact that direct evidence for unification has been so difficult to come by (despite a quarter century of effort and many billions of dollars) demonstrates how high the faith quotient truly is.

The religious undercurrents of the TOE quest are not just subliminal; increasingly, TOE physicists themselves are associating a unified theory with God. The most famous in this camp is Stephen Hawking. In the introduction to Hawking's international bestseller *A Brief History of Time,* Carl Sagan alerts the reader that: "The word God fills these pages. Hawking embarks on a quest to answer Einstein's famous question about whether God had any choice in creating the universe. Hawking is attempting, as he explicitly states, to understand the mind of God." The implication throughout his book is that a unified theory *transcends* space and time and somehow exists "beyond" the realm of material manifestation—a feat traditionally attributed to God alone.

The pages of *A Brief History of Time* are filled with discussions on the options that may or may not have been open to a deity. Hawking's own research into the unification of general relativity and quantum mechanics has led him to a conception of a unified theory which, he says, "has profound implications for the role of God as Creator," and which suggests that a "Creator" may not have had much of a "choice" at all. Hawking's discourse about God does not emulate the humble tone of Copernicus, the stern reverence of Newton, or the ecstatic prose of Kepler. He writes about a deity as if he were telling the reader about a clever older brother, someone whom he admires but whose achievements are well within his own range of comprehension. Hawking has taken the familiarity of Einstein's discourse about a deity even further, so that one is left with the impression at the end of *A Brief History of Time* that Hawking and his god are almost on the same level.

The immense success of *A Brief History of Time*—it has sold more than 5 million copies worldwide—and Hawking's personal success in the public arena, are, I believe, in part attributable to the quasi-religious tone in which he presents the enterprise of contemporary physics. Although his reference to "the mind of God" actually occurs at the very end of the book, it opens the film of the same name. As the filmmakers rightly recognized, in an age when many people are hungering for a rapproachment between the spiritual and the scientific, the concept of the physicist as high priest is immensely appealing. And, like Einstein, Hawking is very convincing in the role. He too has assumed an almost mystical aura, which in his case is compounded by the extreme disjunction between the power of his mind and the lameness of his body. Here he embodies an archetype found in many cultures around the world—the lame or crippled seer. Hawking may be confined to a wheelchair, but his mind soars. Not even many physicists understand his concept of "imaginary time." He is a being seemingly poised at the junction of the human, the subhuman, and the superhuman—and many people long to believe that this disabled physicist might just take us to God.

Ironically, it is Hawking himself who has suggested that his relativistic-quantum cosmology might obviate the need for a "Creator." But he seems to want to have it both ways—at the same time pushing God out of the universe altogether and invoking him as the constant subtext of his work. It is not at all clear from *A Brief History of Time* whether Hawking genuinely believes in a god, or whether he is just indulging in self-aggrandizement. Unlike Copernicus, Kepler, and Newton (and even Einstein in his own way), Hawking is not a serious theological thinker—at least not in his published work. Yet, whatever Hawking's true feelings about God, many people have come to see him as a scientific high priest, the inheritor of Einstein's mantle.

The god Hawking offers us, like the one Einstein offered, is not the spiritual Redeemer of traditional Christianity, but merely a mathematico-material Creator. He is a Pythagorean god stripped of all psychological and ethical qualities, a god whose sole function (if any at all) is to bring into material manifestation a universe based entirely on mathematical "laws." This is the same god the physicist James Jeans invoked when he said, "From the intrinsic evidence of his creation, the Great Architect of the Universe now begins to appear as a pure mathematician." This contemporary god is apparently also guided by strict aesthetic principles, for like the ancient Pythagoreans, he too is said to be obsessed by symmetry. Just as Copernicus and Kepler were convinced that the biblical God would not have created the universe other than by principles of mathematical "perfection," Hawking and his TOE cohorts are committed to a Pythagorean notion of cosmic mathematical "perfection."

If TOE physicists are to be believed, not only will they soon be revealing the "blueprint" for Creation but they will also soon be showing us how the deity realized this cosmic plan—how he turned the equations from an "idea" into a materially manifest universe. Whether they personally believe in a deity or not, many TOE physicists are encouraging the public to associate their elucidation of cosmic evolution with the unfolding of a divine plan. In 1992, when astronomer George Smoot discovered the much sought after ripples in the cosmic microwave background radiation (ripples that are echoes of the big bang), one of his first statements to the press was "It was like seeing the face of God!" Similarly,

Nobel Prize–winning particle physicist Leon Lederman calls a particle known as the Higgs boson "the God Particle." According to current unified theories, the Higgs boson played a crucial role in the original shattering of the superforce.

It is not a little ironic that after almost two centuries of denying cosmic creation, and the godly associations that such an event entrained, physicists are now rushing to embrace both a creative moment and a deity who might accompany it. In recent years they have actually been co-opting God as their own public relations front, for we are increasingly being told that he is the goal of their quest. That is the subliminal message of *A Brief History of Time*. By following the quest for a unified theory, Hawking suggests that we will see into "the mind of God." Leon Lederman has been even more emphatic. His 1993 book, *The God Particle,* was essentially a long argument for why America should fund the now-defunct $10 billion-plus Superconducting Supercollider, a machine whose main purpose was to have been to look for evidence of unification. The primary goal of the Supercollider was to find the Higgs Boson, Lederman's "God Particle." Give us the $10 billion, Lederman seemed to be saying, and we physicists shall deliver the Lord unto thee.

Lederman's book is peppered with passages from something he calls "The Very New Testament." When introducing the Supercollider, he also introduces this fictional text, from which comes the following "quote": "And the Lord came down to see the accelerator which the children of men builded. And the Lord said, Behold the people are unconfounding my confounding. And the Lord sighed and said, Go to, let us go down, and there give them the God Particle so that they may see how beautiful is the universe I have made" (The Very New Testament, 11:1). Throughout his book Lederman refers liberally to God; he even calls one chapter on his own research "How We Violated Parity in a Weekend . . . and Discovered God." The unmistakable implication is that particle physics is a direct path to a deity.

Bolstering his religious subtext, Lederman quotes with approval fellow particle physicist Robert Wilson's association of accelerators with cathedrals. "Both cathedrals and accelerators," he writes, "are built at great expense as a matter of faith. Both provide spiritual uplift, transcendence, and, prayerfully, revelation." Lederman's invocation of accelerators as places of worship, his invention of a pseudo-religious document, and his naming of his own book are all, no doubt, attempts to make what is otherwise a rather dense and technical work more popularly appealing. Whether or not he truly believes the Higgs boson is God's particle is impossible to tell. As with Hawking, it is not at all clear whether Lederman's theologizing comes from a genuine religious faith or scientific hubris—or perhaps just a desire to sell books. What matters is that he has clearly understood that the combination of God and physics is tremendously appealing to the public. Underlying the whole exercise is an assumption that physicists are people who have the credentials to talk about God, that they are people whom the public will accept in a theological role.

Stephen Hawking, Leon Lederman, and George Smoot—these are men at the heart of contemporary physics. Smoot is one of the world's foremost astronomers, Lederman a Nobel laureate, and Hawking one of the foremost theorists of our time. All these men have publicly associated the quest for a unified theory with God. In drawing an association between contemporary physics and God, they are not alone. Indeed, this kind of dialogue has become endemic among physicists—at least as far as their popular writing is concerned. Among others who have written on this theme is English physicist Paul Davies, author of *God and the New Physics* and *The Mind of God*. Going further than anyone, in 1994 general relativity expert Frank Tipler published a book entitled *The Physics of Immortality: Modern Cosmology, God and the Resurrection of the Dead,* in which he claimed to have found "a purely scientific theory for an omnipresent, omniscient, omnipotent evolving personal God."

Some of these men, notably Paul Davies, are genuinely seeking a rapprochement between science and spirituality. Two others are John Polkinghorne, a particle physicist and Anglican priest who is the author of *The Faith of a Physicist,* and Robert John Russell, a physicist and theologian who is the founder of the Berkeley-based Center for Theology and the Natural Sciences. But many physicists using the God drawcard are not engaged in serious theological or spiritual thinking. Following a millennia-old tradition that has associated mathematically based science with divinity, they simply assume it is legitimate to present their activities in a quasi-religious light. Despite the supposedly secular climate of twentieth-century science, some physicists are once again demanding that we see them as high priests, leading humanity "upward" toward transcendent, even divine knowledge of the world.

From Pythagoras to Newton, the search for mathematical relationships in the world around us was conducted within a formalized religious framework. That a religious undercurrent should come bursting out again seems to me not so much a surprise as an inevitability. Like any flesh-and-blood man, Mathematical Man has an inherent psychic inertia. Just as each of us has carried into our adulthood the inculcations of our own childhood and adolescence, so Mathematical Man has carried the resonances of his past. Yet it is precisely in the reemergence of this priestly image of the physicist that I believe we can locate a significant factor inhibiting women's entry into the field. Just as there were no women among the founders of relativity or quantum mechanics, so there are few female names among those searching for a theory of everything. On the eve of the twenty-first century, our mathematical world picture is *still* being constructed largely by men, and women *still* make up only a very small percentage of the inner interpretive core of the physics community.

10

$$\# \quad \varnothing \quad \div \quad \lessgtr \quad < \quad \infty \quad = \quad + \quad \pm \quad \sim \quad \sqrt{\ } \quad >$$

The Ascent of
Mathematical Woman

THROUGHOUT THE TWENTIETH CENTURY, WOMEN HAVE HAD TO CON-
tinue to fight for a place in the community of physics. And they are
still doing so. As nuclear physicist Fay Ajzenberg-Selove remarks in
her recent autobiography, "A giant step forward will occur when
excellent women have no more difficulties than a man would in
obtaining tenured faculty positions in physics departments at research universities." Ajzenberg-Selove has felt the cold wind of
sexual discrimination firsthand. In the 1950s she was told charm-
ingly, but firmly, by the chairman of the Department of Physics at
Harvard that as a woman she would not be eligible for an instruc-
torship. In another episode during that decade she was invited by
colleagues at Princeton's physics department to use the university's
cyclotron for a series of experiments, but she could do so only by
creeping around at night, because the chairman of the department
had a rule: No women in the building.

Such blatant discrimination is no longer sanctioned in the

United States, but covert sexual discrimination in science has by no means disappeared. The Harvard physics department did not accept women as faculty members until the 1970s, when they were forced to do so by antidiscrimination laws, and even then it was only as junior faculty. Not until 1992 did a female physicist gain tenure at Harvard; and as of early 1995, the Princeton physics department had still not given tenure to a woman. No doubt both departments would argue that the available women have simply not been of the first caliber, but as Ajzenberg-Selove points out, "there are plenty of male faculty members at Harvard, or anywhere else, who are second rate." She goes on to remark rather wryly that "I will believe that discrimination against women has stopped when I observe that second-rate women are given tenure." Although her autobiography recounts stories of many men who have helped her during her long and distinguished career as a nuclear physicist (particularly Tom Lauritsen at Caltech and her husband, the particle physicist Walter Selove), it remains the case that the culture of physics as a whole has by no means accepted women as equal partners.

One woman whose life epitomizes the obstacles women have faced getting into physics in the twentieth century is the brilliant Chinese-American particle physicist Chien-Shiung Wu, whom many believe ought to have won a Nobel Prize. Wu was born in China in 1912, a time when it was not uncommon to bind Chinese girls' feet, but, as with many of the women we have seen in this book, her father, Wu Zhongyi, was an extremely enlightened man who believed in equal rights for women. Having quit engineering to participate in the Chinese Revolution of 1911, Zhongyi returned home afterward to the town of Liuhe, where he opened the region's first school for girls. Wu's mother, Fan Fuhua, also worked in the school, and she visited local families to urge parents to stop binding their daughters' feet and give them an education. After a stellar school career, Wu was selected to attend China's elite National Central University, where she majored in physics. But since Chinese universities did not offer graduate education in physics, Wu realized she would have to go abroad to complete her studies. Once again, an enlightened man stepped in to help her—

this time her uncle, who had become rich starting China's first long-distance bus company. He offered to pay for her to go to America.

Wu's original intention on going to the States was to earn her doctorate and return to help modernize her own country. Like Marie Curie, she longed to use science in the service of her people. She had planned to go to the University of Michigan, but, on arriving in America in 1936, she learned that female students there were not allowed to use the Student Union building, so she enrolled at Berkeley instead. There she became part of Emilio Segrè's nuclear physics group and earned her Ph.D. in 1940. After graduation she quickly established a reputation as an expert on nuclear fission (Robert Oppenheimer called her "the authority"), but the Berkeley physics department refused to hire her as a professor. In the early forties not one of the nation's top twenty research universities had a single woman physics professor—and Berkeley was not about to be the first. Segrè, himself a future Nobel Prize winner, was furious at the faculty's refusal to provide a suitable post for Wu, but, at a time when anti-Asian fever was running high, there was little hope of changing anyone's mind.

Because of her expertise in nuclear fission, Wu was recruited into the war effort, and from a base at Columbia University she helped develop sensitive radiation detectors for the Manhattan Project. When the war ended, she was one of the few Manhattan Project physicists invited to stay on at Columbia as a researcher—but still she was not made a faculty member. Not until 1952, when she had established a reputation as one of the best experimental nuclear physicists in the world, was she offered even an associate professorship. She remained at Columbia until her retirement in 1981. Wu's capacity for work, her obsession with excellence, and her "slave-driver" approach to her graduate students earned her the stereotypical nickname Dragon Lady, but this tiny and delicate woman, who always wore traditional Chinese silk dresses, performed difficult experiments with unrivaled precision.

In 1956 it was precisely this quality that was called for when two young Chinese-American physicists put forward an astounding proposition. Tsung-dao Lee and Chen Ning Yang proposed that in certain particle reactions a basic law of physics might be broken. This was the law of parity. According to this law, particle reactions should always be symmetrical, but Lee and Yang suggested that in certain reactions asymmetry *might* creep in—and they urged experimentalists to put the cherished belief to the test. Most physicists, including Lee and Yang, were convinced, however, that the parity law would hold up to scrutiny, and since the relevant experiments were so difficult no one was willing to waste their time "proving" something everybody already knew. Yet Wu thought the problem sufficiently important to take seriously: If nothing else, doing the experiments would confirm the validity of symmetry once and for all. She quickly assembled a team, based at the National Bureau of Standards, and drove them hard. In early 1957, they had evidence that, contrary to all expectation, particle reactions were not always symmetrical. Parity was sometimes violated. The news sent shock waves through the physics community. Not only had Wu's team disproven a supposedly fundamental law of nature, but their results had important implications for the understanding of the weak nuclear force. In swift recognition of this discovery, just ten months later Lee and Yang were awarded the Nobel Prize for physics.

Lee and Yang alone. Sharon Bertsch McGrayne, author of *Nobel Prize Women in Science,* has noted that "the Nobel Committee has made similar rulings before," rulings in which the theorists alone have won the prize, not the experimentalists who proved their ideas. Yet in this case the experiment was everything, for even the theorists themselves had expected the *opposite* result. Unfortunately for Wu, after rumors of her promising results began to leak out, others rushed into the fray and a team led by Leon Lederman, using more sophisticated equipment, produced similar results, which they published just a few days before her formal announcement. Thus while Wu had been the first to get any results for parity violation, and while she had been the first to think the experiments worth doing at all, she had been scooped. Although Wu lost the Nobel, she received every other major physics prize, including the Wolf Prize and the prestigious Comstock Award from the National Academy of Sciences. She was also the first woman to be made president of the American Physical Society—and in the wake of her success was finally made a full professor at Columbia. Still, there is nothing as mythical as a Nobel, and Wu was sorely disappointed not to have won it.

Only one woman other than Marie Curie has *ever* been awarded a Nobel Prize for physics, the German-American theorist Maria Goeppert-Mayer, who in 1963 shared the award with Hans Jensen and Eugene Wigner for her work on the shell theory of the atomic nucleus. Born and educated in Germany, Maria Goeppert (1906–1972) moved to the States in 1930, after falling in love with and marrying a young American physical chemist named Joseph Mayer. Despite a brilliant doctoral thesis on quantum mechanics, she too found that no American university would employ her as a professor, and so, like Emmy Noether, she pursued her work as an unpaid, unofficial member of various academic departments. In her case this lasted for almost thirty years. Not until 1960, a decade after she had completed her Nobel Prize–winning work, was this first-class theorist offered a full-time paid academic job. Until then she had to be content with honorary positions and a supportive husband. Fortunately, Maria Goeppert had found a rare partner indeed. Although Joe Mayer worked in a different area from his wife, from the start he recognized in her a superior talent, and throughout their marriage he pushed her to continue her own work. It is no coincidence that the only two women ever to win Nobels in physics were married to singularly enlightened and supportive men.

Conditions for women in physics, indeed for women in all sciences, have improved tremendously since the 1970s. Nonetheless, in this unprecedented age of feminist consciousness, physics remains overwhelmingly male-dominated. Although there are now many more women physicists than ever before, and although they are now active in every branch of physics, it remains the science with by far the lowest participation by women. According to the American Institute of Physics, in 1992 (the latest year for which full figures are available) women in the United States received only 15 percent of bachelor's degrees in physics, 11 percent of Ph.D.'s, and they constituted just 3 percent of full professors in the field! In 1994 women still held only 3 percent of full professorships, but they also held 8 percent of associate professorships and 10 percent of assistant professorships. The fact that women *are* slowly gaining a greater percentage of doctorates and junior academic positions indicates that their representation in the field is likely to be better in the future. But while this is certainly cause for hope, the rate of growth is slowing down, and we cannot be complacent.

The poor percentage of women in physics is brought into stark relief when we compare physics with other sciences. While in 1991 women earned just 15 percent of physics bachelor's degrees in the United States, they earned 40 percent of those in chemistry, 47 percent in mathematics and statistics, and 51 percent in life sciences. Similarly, while in 1992 women earned 11 percent of Ph.D.'s in physics and astronomy, they earned 19 percent of those in mathematics, 23 percent in environmental science, 26 percent in chemistry, 39 percent in biological and life sciences, and 47 percent in social sciences. In the past few decades, women have made enormous advances in the social, biological, and life sciences, yet they have not broken into physics in anything like equitable numbers. As we approach the twenty-first century, women remain disproportionately clustered in the so-called soft sciences. Why, we must ask, is this the case? Why is it that of all the sciences, physics remains by far the most male-dominated?

Part of the answer lies in the barriers women still face gaining entry to any natural science, with physics being an extreme example of the general case. In *The Outer Circle: Women in the Scientific Community,* sociologist Harriet Zuckerman and her coauthors have written that, despite the gains women have made during this century, "science remains dominated by men, not only numerically, but in the exercise of authority, power, and influence." No fact brings this home more starkly than the overall statistics for Nobel science prizes. Since they were instituted in 1901, just over 400 men have won science Nobels. (There are three categories: physics, chemistry, and physiology or medicine, and in each year up to three people may be awarded the prize in each category.) Along with these 400 men there have been just 9 women. You can count them on your fingers. While we are here, let us acknowledge their achievements, for if Ginger Rogers had to do everything Fred

Astaire did but backwards and in high heels, these women had to do everything their male colleagues did but they may as well have been doing it backwards, in high heels, blindfolded, and up a steep slope. They are: Marie Sklodowska Curie, Physics, 1903, Chemistry 1911; Irène Joliot-Curie (her daughter), Chemistry 1935; Gerty Radnitz Cori, Medicine 1947; Maria Goeppert-Mayer, Physics 1963; Dorothy Crowfoot Hodgkin, Chemistry 1964; Rosalyn Sussman Yalow, Medicine 1977; Barbara McClintock, Medicine 1983; Rita Levi-Montalcini, Medicine 1986; and Gertrude B. Elion, Medicine 1988.

The small number of women who have been awarded science Nobels highlights one problem other women face with respect to becoming scientists: the lack of many role models. To this day, Marie Curie remains the only female scientist most people can name. History, as it is generally taught, suggests that science—particularly the "hard" sciences—has always been a male pursuit, and that view is strongly reinforced by popular culture. In films, on television, in comic books and even music videos, mathematico-scientific "brains" are almost always male. Where are the female equivalents of Mr. Spock or Data from *Star Trek*, or Dr. Who from the English television series? While there have been several Hollywood films about boy geniuses with an aptitude for math, most recently *Little Man Tate* and *Searching for Bobby Fischer*, I know of none about a girl mathematical genius.

The message given to girls in class is often little better. In *Failing at Fairness: How America's Schools Cheat Girls*, a superb if depressing study of gender inequity in American education, researchers Myra and David Sadker have documented how teachers tend to encourage boys in math and science classes far more than they do girls. This is independent of the teacher's sex and has also been found in Australian and European studies. In classrooms across the nation, the Sadkers describe how teachers choose boys to answer questions more often than girls, how they give boys longer to speak, and give them more feedback about their answers. Although such inequity is by no means confined to math and science classes, it becomes especially acute in these subjects. As a paradigmatic example, the Sadkers describe an incident in an elementary school where a teacher ordered a group of girls away from the math blocks so the boys could get in and do their work. Such incidents, they report, are common at all levels of education.

In a 1992 survey by *Glamour* magazine, 74 percent of those responding said they had a "teacher who was biased against females or paid more attention to the boys." Math was selected as the class in which inequity was most likely to occur. In the hectic classroom environment, boys talk longer, interrupt more, and commandeer more equipment. While not all boys are gung-ho, nor all girls wallflowers, the school math and science environment is not experienced equally by both sexes. As the Sadkers have demonstrated, boys get more time, more attention, and most important, more encouragement than girls.

After school, women all too often continue to face tremendous barriers in science. As Zuckerman and other analysts of contemporary scientific culture have noted, much of the important activity in science takes place informally, so that even when women are in the system they often remain outsiders to vital informal networks. In her ethnographic study of the particle physics community, anthropologist Sharon Traweek has described how in Japan it is not socially acceptable for the women physicists to go out to dinner with their male colleagues, yet this is precisely where important ideas are often discussed. Although American society is more relaxed about male-female dining, "old boy" networks still exist in the scientific community. In her autobiography, Ajzenberg-Selove notes that women physicists are still less likely to be asked to give colloquia talks than their male colleagues; Zuckerman et al. report that women scientists are less likely to get preprints of papers before they appear in journals. Thus, they are more likely to learn the "hot" news in their field later than their male colleagues, and, because timeliness is everything in science, this can be a distinct disadvantage.

Men also dominate the editorial boards of most science magazines, and there is some evidence that women are less likely to have their papers accepted. Most important, women have a harder time finding mentors among male faculty members. Because mentors often play a major role in launching a young scientist's career, this is a critical issue. Women do tend to get promoted if they stay in the field, but on average this takes several years longer than for their male counterparts. They also tend to have fewer assistants, less equipment, and less actual space to work in. Furthermore, outright discrimination has by no means disappeared. In one study senior university scientists were sent fictitious curriculum vitae—some under the name Joan and some under John—and were asked to make a recommendation for appointment. Often they recommended John for an associate professorship but Joan for only an assistant professorship, even when the CVs were exactly the same.

Yet in spite of all these obstacles women *are* making major progress in most sciences. Why not in physics? I find it extraordinary that in the last thirty years we have seen women become heads of government in India, the United Kingdom, Norway, Canada, and even Pakistan and Turkey, yet in that time not a single woman has won a Nobel Prize in physics. By comparison, in the same period four women have won Nobel prizes for medicine. If women can now rise to the top of government in two Muslim nations, why are they not rising to the top in physics? Why is it that the science that so fundamentally shapes the Western world picture is still so male-dominated?

As other commentators have noted, one factor that undoubtedly continues to impede women's participation in physics is the enduring belief that women are biologically inferior to men with respect to mathematical ability. Although there is now general agreement that women are just as intelligent as men, there is a widespread belief that women's brains are more naturally attuned to linguistic skills, whereas men's are more naturally attuned to mathematical skills. Because physics is the most mathematical of sciences, this view reinforces the traditional cultural stereotype that it is a man's subject. During the last decade, male-female brain difference has once again become a hot research topic, and a number of studies have purported to show evidence of gender disparity in mathematical ability. It is not just men, but also women, who are doing this research. One of the leaders in the field is Canadian psychologist Doreen Kimura, who has reported that men perform better on tests requiring "mathematical reasoning" and those requiring an ability to analyze complex spatial relationships.

The conclusion Kimura and others have drawn from such studies is that men are innately better at the kind of thinking required for excellence in mathematics. Women's compensation is that they are supposedly better at "perceptual" and verbal skills. This bipolarity between the perceptual-verbal female mind and the analytical-mathematical male mind is, in effect, an updated version of the theory of complementarity so popular during the eighteenth century. Once again, men and women are said to be essentially equal, but each gender is said to have its own unique sphere of excellence, with mathematics falling "naturally" into men's domain.

At first glance this argument appears to have much validity, but one of the problems that emerges on closer scrutiny is that as children boys and girls exhibit no difference in performance on formal assessments of mathematical ability. Since the late 1960s, the United States government-funded National Assessment of Educational Progress (NAEP) has been conducting surveys of American student proficiency in key areas, including mathematics and science. Surveys are conducted on students at age nine, thirteen, and seventeen. According to 1990 NAEP figures, at age nine, out of a possible 500 points, girls' mean math proficiency score was 230.2, whereas boys' was 229.1. As a report by the National Science Foundation stresses: "Levels of proficiency for males and females in this age group are remarkably similar." The same is also true at age thirteen.

By age seventeen, however, a slight difference has developed, and seventeen-year-old girls lag boys by 1 percent. But this small discrepancy does not tell the full story, because although girls' *average* is only a little below boys', by the end of school, far fewer girls are rating in the top percentile bracket of the NAEP math assessment, and fewer are taking advanced math courses. Furthermore, on the all-important Scholastic Assessment Test or SAT (required for admission to most colleges), males outscore females by 50 points on the math section. Nonetheless, when we look at tertiary education, we once again find that at the junior level there is not a major difference between men and women. In the United States, just under half of all bachelor's degrees in mathematics and statistics are now earned by women. It is only when one looks at higher degrees, particularly to Ph.D.'s, that the percentage of women drops off dramatically.

What both NAEP figures and bachelor's degree statistics suggest is that females are *not* innately inferior to males in mathematical ability but that as women move through the educational system, their mathematical talents are not developed as fully as those of their male counterparts. This view is also supported by the Sadkers' research. Time and again they have found that when girls reach puberty they begin to play down their intelligence—particularly with respect to math and science. Being a "brain" is all too often seen by teenagers of both sexes as unfeminine, and, rather than risk this judgment, many bright girls take the more socially acceptable path of pretending to be less intelligent than they are.

The Sadkers also include the story of one young woman who refused to hide her mind. In 1991 Ashley Reiter was the national first-place winner in the prestigious Westinghouse Talent Search, for her research in mathematics. As part of the prize, organizers arrange for winners to interview someone of their own choosing. Most of these young superstars ask to meet prominent scientists, but Reiter asked to meet the Sadkers. She told them how, in refusing to play dumb, she had found herself a social outcast at school. She was relieved to learn that her experience was all too common.

Even the "scientific" evidence for female mathematical inferiority has been challenged. In *Myths of Gender,* Anne Fausto-Sterling, professor of biology and medicine at Brown University, has analyzed many of the relevant studies and concluded that most do not stand up under rigorous scrutiny. Fausto-Sterling does not suggest that Kimura and her cohorts fudge their results, rather that their statistical methods are often open to interpretation. As she points out, if one *wants* to find a difference between two groups of people, and one tries enough methods of comparisons, then a difference *can* usually be found. Yet, as Fausto-Sterling stresses, the real issue is what conclusions can validly be drawn from such statistics. Do they really indicate an innate biological difference in men's and women's mathematical ability?

According to Fausto-Sterling, and a number of other researchers who have looked at this issue, there is *no* clear evidence for an innate gender difference in mathematical ability. In all the studies that do find some genuine difference, it is invariably *very small*—no more than a few percentage points. Fausto-Sterling argues that this differential can easily be explained by differences in socialization rather than biology. For instance, boys are more likely than girls to have toys like Lego that help teach the young mind about complex spatial relationships. Similar informal mathematical training extends throughout boys' lives in activities such as woodwork and metalwork, and in sports such as baseball and basketball. By contrast, many girls' play activities do not provide much in the way of informal mathematical training. Thus, Fausto-Sterling argues that when adult women perform slightly worse on tests of spatial perception it does not necessarily indicate an *innate* lack of mathematical ability. The cause, she says, is not a deficiency of the relevant neuronal structures but simply a deficiency of experience.

That biology is *not* the explanation for women's low representation in physics is also suggested by a comparison of the number of women in physics versus those in mathematics itself. At all levels women are far better represented in mathematics than in physics. As already noted they now receive 47 percent of bachelor's degrees in mathematics and statistics, but only 15 percent of those in physics. For Ph.D.'s the figures are 19 percent for math and stats, and 11 percent for physics. This is not a new pattern. Throughout the period American women have had access to higher education, they have always been better represented in math. In the 1921 who's who of American science, titled, tellingly, *American Men of Science,* there were 42 women mathematicians listed but only 21 female physicists. The 1938 edition listed 151 women mathematicians but only 63 physicists. In sheer numbers there have consistently been more women in math than in physics—although for men the opposite has been true. Looking at percentages, the difference becomes even more stark. Throughout the twentieth century, women's percentage representation in math has been *at least twice* what it has been in physics.

By almost any measure, women have found it a good deal harder to break into physics than into mathematics. Again, this suggests not an innate female deficiency but a powerful cultural inertia that continues to militate against women physicists. The "old boy" networks, the tacit double standards that continue to operate in many universities, and the lack of many female physicists as mentors for young women—all these are important factors that contribute to gender inequity in physics. But in an age when women represent almost half of all American biological and social scientists, and over a quarter of chemists and mathematicians, their acute underrepresentation in physics demands a further explanation.

It is one of the premises of this book that the historic problem women have faced trying to break into science parallels the problem they have faced trying to break into the clergy. Women have had to fight on the one hand for the right to interpret the "book" of Nature, and, on the other hand, for the right to interpret the books of Scripture. On both fronts it has been a long struggle, yet, just as women are now gaining the right to be ministers in many denominations of the Christian church, they are making headway in many "denominations" of the church scientific. In this analogy the problem women face trying to break into physics parallels the problem they face trying to break into the Roman Catholic clergy. Both are the most entrenched bastions of male power—one in science, the other in Christianity—and as such both are proving the last to accede to female infiltration.

This analogy is not simply a colorful metaphor, for as I hope I have shown, modern physics has been intimately intertwined with Christianity for most of its history. Such a long association cannot be expected to dissolve easily—and as we have seen, physics is still imbued with powerful resonances of this past entanglement. Just as there continues to be an immense cultural inertia behind the male-only Catholic priesthood, so there continues to be a powerful inertia behind the idea of a male-only scientific "priesthood."

In particular, I suggest, the cultural inertia inhibiting women's advancement in physics stems from the male-female heaven-earth dichotomy that is still strongly embedded in the Western subconscious. Ever since the Homeric era, women have been cast on the side of the material, the bodily, and the "earthly," while men have been cast on the side of the spiritual, the intellectual, and the "heavenly." For most of the Greeks, particularly Aristotle and his followers, it was men alone who could aspire to psychic transcendence, whereas women with their supposedly defective souls were said to be forever trapped in the material prisons of their bodies. We have come a long way since Aristotle, but Western society continues to expect that women will remain "grounded" in the physical, the personal, and the domestic. The ancient association of maleness with psychic transcendence continues to underpin the male dominance of mathematically based science today.

This "grounding" of women is precisely what has often allowed male physicists the freedom to devote themselves to their quest for "transcendence." One reason Einstein could be so disdainful of the personal and material was that in the latter half of his life he had a second wife who devoted herself to his care. While he was off pursuing cosmic harmonies, Elsa cooked, cleaned, kept house, and created around him a zone of domestic comfort and calm. It is a pattern not uncommon in physics today. In Traweek's study of the particle physics community, she has noted that 95 percent of this community are men, and the ethos of the community puts great stress on stable marriages wherein wives stay home and care for hearth and husband. As with Elsa Einstein, Traweek reports that almost all the wives she interviewed felt honored to be playing this supportive role to men at the forefront of such important scientific knowledge. The few women particle physicists tended to be married to other particle physicists, or at least to other physicists.

The expectation that women *will* remain "grounded" pervades Western culture, and affects almost every facet of women's lives—including their intellectual lives. From an early age, girls are encouraged to be concerned about their bodies, their appearance, and domestic order; in other words, to be deeply concerned with the *material*. While young men are by no means free from these concerns, they do not dominate boys' lives to the degree they do for many girls. The quest for "cosmic harmonies" is a quest for something utterly *disembodied*, something utterly *immaterial*; as such, it is the antithesis of what girls are generally taught to find important. Modern Western society does not have a model of female intellectual transcendence. In the Christian West, the tradition of intellectual transcendence has always been associated with a male priesthood. Given this long history, it is not surprising that when women *did* break into science, they would be concentrated in the "earthly" life sciences, and that the "heavenly" mathematically based science would remain the last arena of male hegemony.

Having documented the long struggle Mathematical Woman has faced, the question that confronts us is: What does it matter if women participate in physics or not? What does it matter if this science is done primarily by one sex or the other? Why should we care? There are several major reasons that women's low rate of participation in physics warrants our concern. First, left to himself Mathematical Man has developed what I believe is an untenable world picture. Second, some of the goals physicists are now setting for themselves have disturbing consequences for society at large, and I suggest that these are goals women would be considerably less inclined to pursue. Nowhere is this clearer than with the quest for a unified theory of the forces. Before we consider the general issue of physicists' world picture, I would first like to look at their obsession with a so-called theory of everything.

Many of those involved in this quest have now adopted an attitude that this is something society must fund—regardless of the cost. Thus, it becomes a matter for us all because we would have to supply the money via taxes. In itself, a unified theory of the forces and particles is a very intriguing goal. While it is definitely overstating the case to call it a theory of *everything*, a single theory that would unite forces, particles, space, and time is certainly worth considering. At the same time, it has now become clear that progress toward a unified theory is going to cost billions of dollars, for that is what it takes to build the accelerators unified theorists need to test their theories. In a world reeling with problems of pollution, overpopulation, starvation, land degradation, deforestation, and loss of biodiversity, we cannot avoid asking: Is it responsible to spend billions of dollars looking for a theory that, no matter how beautiful, is unlikely to have *any* application to daily human life? Furthermore, we must ask ourselves if this is the way we ought to be spending our precious science dollars.

The cost of pursuing a unified theory can be gauged by the now-defunct Superconducting Supercollider (SSC) which U.S. physicists had begun building beneath the Texas plains. This was the machine they had hoped would reveal new evidence for unification. Although the SSC was originally budgeted at around $2 billion, by mid-1993 it had spiraled to over $10 billion, and some pundits were predicting it may cost as much as $13 billion to complete. At that point Congress pulled the plug. It was a serious setback to the TOE community, but their dreams are by no means shattered. There is every chance the European community will build its own super-accelerator, the proposed Large Hadron Collider (LHC). As I write this, the relevant governments are haggling over the funding arrangements. The price for the LHC is more modest, just over one billion pounds (around $1.5 billion U.S.), and although it will not be as powerful as the SSC, there is hope that it too will open new windows onto the realm of unification. American particle physicists have not given up the dream of their own super-accelerator. New technologies are being developed, and new approaches to politicians will no doubt be made in the future. But whatever new methods are developed, it seems unlikely that serious progress toward a unified theory can be made without spending a great deal of money.

In the face of rising taxpayer resistance, champions of a unified theory have resorted to increasingly troublesome tactics to muster public support. In a 1993 Op-Ed piece in *The New York Times,* particle physicist Leon Lederman wrote that if Congress abandoned the SSC, "something of immeasurable importance would vanish from the scientific enterprise." "It is easy," he continued, "to vote for the immediacy of saving money. It is more difficult to fight for the kind of nation we want for our grandchildren: wealthier, wiser, and in harmony with a universe we understand." Robert

Wilson had used a similar tactic to argue the case for the Fermilab accelerator. When asked by a senator if Fermilab would have any relevance for the security of America, Wilson replied that it would serve no such purpose. Then he went on: "It has only to do with the respect with which we regard one another, the dignity of man, our love of culture. It has to do with, are we good painters, good sculptors, great poets? I mean all the things we really venerate and honor in our country and are patriotic about. It has nothing to do with the defense of our nation, only to do with making it worth defending." The implication in both cases was that doing particle physics and pursuing a unified theory are necessary for the continued integrity of American culture. And as if even that might not be sufficient inducement, TOE champions have increasingly been implying that the quest for a unified theory will also take us to God.

Given that even the most ardent TOE champions acknowledge that a unified theory is unlikely to have any practical applications in day-to-day life, even for military purposes, it is sheer self-indulgence to ask society to spend billions of dollars on this quest. If a unified theory could be achieved at a low cost I would be the first to vote it into the budget, for ever since I learned about relativity and quantum mechanics I have dreamed I would live to see them united. But in the face of such vast expenditure, at a time when the world faces so many problems that science *can* help us resolve, I believe it would be an act of gross social negligence to pour further billions into the quest. No knowledge, for its own sake, is worth this price. Infinite lust for knowledge, like infinite lust itself, is unjustifiable. It is the overweening lust for knowledge that is the real sin referred to in the biblical tale of the Garden of Eden. Though I personally would be fascinated to see a unified theory, it is *not* society's duty to pay for it.

In their demands that society should keep paying, TOE physicists have become like a decadent priesthood, expecting the populace to build them ever more lavish and costly cathedrals, with spires reaching ever higher. Like the medieval Scholastics, they *are* pondering deep and intricate questions about the world, but the deeper they go on their chosen path and the more intricate their questions become, the more they are venturing into territory that is not only irrelevant to most people's lives, it is beyond our comprehension. Yet they seem to expect that the rest of us should be happy to be spectators to their moments of revelation, their stages of enlightenment. They seem to think we should feel honored to keep on funding ever higher stairways into their idea of heaven, so that one day we can watch them ascend into it. The fact that most of us will not be able to follow seems not to concern them in the least. They will have found the "light," they say, and the rest of us will be immeasurably enriched—even if we cannot see it for ourselves.

I believe we need a new culture within the physics community, one that is not so fixated on quasi-religious, highly abstract goals and that is more concerned with contributing positively to the needs and concerns of society at large. I suggest that one change women might precipitate is to move the physics community in this direction. While it is not true that all women are caring, sharing, nurturing beings concerned with social welfare, or that all men are megalomaniacs reaching for the "heavens," I believe that equitable participation by women *would* help to shift the focus of physics away from the present obsession with a unified theory. One reason for this belief is that women are not so acculturated to seek this kind of "transcendence" in the first place. What I am suggesting is that physics itself needs to be more "grounded"—a quality our society tends to acculturate into women.

The desire for a unified theory is, at heart, a desire for something *beyond* nature—for a set of equations that transcend the inherent mortality and mutability that is the hallmark of all material, or "natural," phenomena. The ancient Pythagoreans had believed that numbers were timeless, immutable, disembodied blueprints for material form, and a theory of everything is contemporary physicists' version of this idea. I believe we need a culture of physics that embraces, rather than tries to escape from, nature; a culture that values the *embodied*, and is not so obsessed with escaping into disembodied abstraction. As I noted at the start of this book, the problem with physics is not that its practitioners use mathematics to describe the world but how they use it and to what goals they apply it. There is nothing in a mathematical approach to nature that demands a fixation on a unified theory or on "transcendent" abstractions. Like all sciences, physics could have different goals—and women could play an important role in helping to shape new ideals for this science.

The second reason we need to encourage more women into physics is intimately entwined with the first. The very reason TOE physicists have become so obsessed with subatomic particles and forces is that they have created a hierarchical world picture in which these things reside at the apex of a pyramid of value. During the last four centuries, physicists have evolved a view of nature in which the more mathematically reducible an entity, the more "fundamental" and, by implication, the more important it is regarded as being. Particles and forces are so highly valued because they are the most mathematically reducible entities in the pantheon of science, and so they have been viewed as the most important entities. It is this hierarchical world picture that needs to be challenged—and which I believe women might help to dismantle.

Philosophers of science have pointed out that physicists' hierarchy of fundamentalness is not written in nature but is a social construction. Although physicists often assert that their science discovers purely objective knowledge of the world, the notion of pure objectivity is a myth. Science is done by subjective people, not by objective "its," and all people operate within a cultural milieu that ineluctably colors their thinking. There simply isn't any such thing as culturally neutral science. Our scientific pictures of nature are products not only of empirical discovery, but also of human biases and culturally influenced ways of thinking. As we have seen, physicists' hierarchical world picture originally emerged during the social turmoil of the seventeenth century, and this new worldview was deeply influenced by the desire for a conception of nature that could serve as a model for a hierarchical and patriarchal social order.

Furthermore, with the recent emergence of complexity theory (an outgrowth of chaos theory), physicists' hierarchy of "fundamentalness" is being challenged by other scientists, including some physicists. One of the important discoveries to come out of complexity theory is that, at every level, nature exhibits behavior that

does not seem to be predictable from the behavior of the constitute parts. Living beings, for instance, are made up of atoms, but
they exhibit behavior that is not written into atoms themselves.
Some champions of complexity, such as physicist Paul Davies, now
argue that at *every* level of nature there are phenomena and laws
which must be seen as truly fundamental.

Along with many feminist philosophers of science, I believe we
need a world picture that is not so fixated on hierarchy, a world
picture in which, as Evelyn Fox Keller has suggested, "disease
would be seen as just as fundamental as subatomic particles"—and,
therefore, just as worthy of vast expenditure. I suggest that women
physicists might play a role in helping to evolve a less hierarchical
world picture. Again, the point is not that women are innately less
hierarchical but that having generally been acculturated differently
than men, they have a somewhat different pool of perspectives to
bring to the endeavor.

Just what these perspectives might be is very difficult to predict,
and many women physicists feel uncomfortable talking about gender difference. Outsiders already, many are keen to stress their
similarity to rather than their difference from their male colleagues.
One who has written about difference is Karen Barad, a theoretical
physicist at Pomona College in California. Barad notes that precisely because women are outsiders in physics they tend to reflect
more on their own role in the community. This self-reflectivity, she
suggests, tends to spill over into reflecting on the subject itself, and
perhaps to a greater willingness to question. For example, she has
found that her work on gender has led her to teach quantum mechanics in a different way. Barad also notes anecdotally that in her
experience women in physics often tend to ask different kinds of
questions, a phenomenon she often sees with nonwhite men as
well. Since the seventeenth century, physics has almost exclusively
been a pursuit of upper-class white males; thus, it stands to gain
fresh insights and perspectives from most other cultural groups.

Greater cultural diversity in the physics community could lead
to new philosophical insights and also to new *science*. One area of
physics in which women are already better represented is astro-
physics, and since the 1960s there have been a significant number

of women who have made major contributions to this field—notably, Vera Rubin, one of the first to discover evidence for cosmic dark matter, and Sandra Faber, whose research group discovered that the great cluster of galaxies to which the Milky Way belongs is moving sideways through the universe "like huge flocks of birds." Other notables include Beatrice Tinsley, Margaret Geller, and Neta Bahcall. In a 1994 interview in *Scientific American,* leading astrophysicist Jeremiah Ostriker commented that the reason he felt so many women in the field had made major discoveries was precisely that they were outsiders. As such, said Ostriker, they simply didn't know that certain things were supposed to be impossible.

The notion that significant numbers of women can affect the prevailing culture of a science, and therefore also its content, is not hypothetical; women are starting to do just that in the biological sciences. Particularly since the 1970s, women have been challenging many of the paradigms of the biological sciences and have been a catalyst for introducing new ways of seeing the organic world. There is a large literature now devoted to this subject and I can mention only a few of the highlights.

One of the most important changes women have wrought in the biological sciences is that across the spectrum, from the level of cells to the level of the biosphere as a whole, women scientists have focused on cooperation rather than competition among organisms. Ever since Darwin's *Origin of the Species,* biologists have stressed competition as the driving mechanism in both evolution and individual behavior. But women scientists have begun to show that cooperation is often just as important. World-renowned cell biologist Lynn Margulis believes it has been the *major* factor driving evolution. Margulis points out, for example, that the very cells of which animal bodies are composed are an evolutionary adaption of several primitive kinds of cells working together symbiotically.

Women have also revolutionized our understanding of the great apes. Dian Fossey and others have shown that our evolutionary cousins are far more cooperative and social than male primatologists had given them credit for. The importance of this discovery lies in the fact that what scientists think about apes has often been used as a model for early humanoids. In other words, scientific understanding of apes today colors our view of human evolution, an area to which women have brought many new perspectives. Until recently, evolutionary theorists tended to view women as passive bystanders in a process driven by the exploits of "Man the Hunter." But now anthropologists such as Adrienne Zihlman are showing that this androcentric view is unjustified. Some even see the reverse, an evolutionary chain driven by the exploits of females. Women scientists are also rewriting the science of their own bodies. Their new perspectives on female physiology have profound implications for the treatment of diseases like breast cancer and cervical cancer, and for the study of pregnancy, menstruation, and menopause. Finally, any list of how women have affected biological science must mention the work of Nobel Laureate Barbara McClintock. Going against the grain of classic genetic theory, McClintock declared that the genetic code of an organism is not a static blueprint that can be read off like a book, but a flexible, dynamic code responding to the surrounding environment. Her notion of a dynamic genome and of "jumping genes" has revolutionized genetics.

McClintock herself has said she came to her ideas by "listening" to her corn plants and trying to see the world from *their* perspective. Evelyn Fox Keller has pointed out that in this respect McClintock was prepared to adopt a subjective approach to her subject, an approach which has traditionally been regarded as feminine. By eschewing the officially sanctioned methodology of science, she made a major breakthrough. Just as McClintock "heard" things in the symphony of genetics which had eluded her male colleagues, women physicists may "hear" things in the mathematical symphony of nature that male physicists have not previously noticed—or have simply not been interested in exploring. If women have brought new insights to genetics and cellular biology, we cannot easily dismiss the idea that they might also bring new insights to physics.

Above all, since women have come into the field they have altered the *culture* of biological science so that not only women, but also men, seek to understand the organic realm in new ways. James Lovelock, the architect of the Gaia theory of the earth, is an excellent example. A colleague of Margulis, Lovelock has developed a view of the entire biosphere as a single holistic organism, which he calls Gaia. The Gaia theory proposes that not only all animals and plants, but also the atmosphere, the oceans, and the soil, are bound together in complex webs of interdependency. This way of seeing has led to important developments in the scientific understanding of chemical cycles in the soil, water, and atmosphere. The point is that when significant numbers of women participate in a science they alter the intellectual climate so that some practitioners of *both* sexes are enabled to see in new ways.

Just as this is already happening in biology, I believe a greater female presence in physics would make a difference. That is not to suggest every woman physicist is, or should be, a catalyst for change. Many women in the field will, like many of their male colleagues, fit into the general stream. Whether male or female, most practitioners in any field go with the flow. Indeed, it is important to recognize that women physicists should not be saddled with a burden to change the culture. What I believe will happen is essentially the same process that has been observed in both the biological and social sciences: When women do break into physics in significant numbers, the climate will simply change—in ways nobody could have predicted.

Since society at large stands to benefit from a greater participation by women in physics, it behooves us all to take seriously the question of how to encourage more women to enter the field. First and foremost, there is a tremendous need for affirmative action at all levels of education. As researchers such as the Sadkers have documented, in far too many schools across the nation girls get a second-rate education in math and science. Effort must be put into training teachers to be aware of bias against girls, and to help them develop strategies for giving equal time to students of both sexes. Similarly, since research has also shown that many girls like to learn in different ways than boys—for instance, they express a greater liking for group activity—strategies must be developed for teaching math and science in ways more attractive to them. Girls at school also need exposure to role models, both current and historical.

Such affirmative action needs to be carried into colleges and universities. One example already in place is Nina Bayer's course in nuclear physics at UCLA. Bayer teaches nuclear physics by looking at the work of the many women who have been in the field, such as Lisa Meitner and Maria Goeppert-Mayer. Most important, young women physicists need access to female professors who can serve as mentors. Of course, not all women feel this need, but many certainly do. The need to actively support and encourage women in science (particularly in the "hard sciences") has been widely recognized, and some schools are beginning to implement programs to do this. One promising model is the WISE Institute (for Women in the Sciences and Engineering), established in 1994 at Penn State. As well as providing a support network for female science students and faculty at Penn State, this multipurpose institute examines ways to draw women into science and raises local, national, and international awareness about the obstacles confronting women scientists. National organizations such as American Women in Science and Graduate Women in Science are also engaged in affirmative action programs.

Such programs *are* making a difference, and the percentage of women in physics *is* slowly rising. Yet we cannot feel certain that affirmative action alone will solve the problem of inequity in physics, because one of the major problems women face is that many find the present culture of physics so alienating they would not want to be part of this culture, even if it did welcome them. I myself left the field because I found the prevailing culture untenable—and I *love* the science. Thus we have a catch-22 situation: We need to get more women into physics so we can create a culture of physics in which more women would *want* to participate.

This is a profound dilemma that philosophers and sociologists of science have only recently begun to understand. In the past there was a general belief among champions of women in science that if only they were given equal opportunities, the problem of inequity would soon be resolved. Now, however, there is a growing realization that, as science historian Londa Schiebinger has put it, "the assimilationist model doesn't work." To put it bluntly: It is not just a matter of helping women to change so they will be comfortable with the culture of physics, we also need to consciously work on changing that culture itself.

The essence of the problem, as Keller has discerned, is that over the last four centuries Western culture has evolved conceptions of "science" and "femininity" as polar opposites. She writes: "If science has come to mean objectivity, reason, dispassion, and power, femininity has come to mean everything that science is not: subjectivity, feeling, passion, and impotence." This polarization is especially acute in the "hard" science of physics. Many women feel that in order to become physicists they would have to reject the very characteristics regarded as feminine, that the price of success would entail effacing their "womanliness." Such a fear is by no means ungrounded. A prominent physicist once declared "only blunt, bright bastards make it in this field." Since the prevailing conception of femininity is the opposite of the "blunt, bright bastard," there is a genuine problem here. Keller has rightly stressed that as long as we live in a culture in which success in science is seen to be based on qualities considered unfeminine, we shall never achieve parity in science—no matter how many affirmative action programs we have.

What, then, is the solution? For a start, Keller notes, we must have a more pluralistic notion of success in science. "Blunt, bright bastards," she suggests, are *not* the only people who can be good at science; as counterexamples we might cite Michael Faraday and James Clerk Maxwell. We must recognize this fact and seek to encourage different sorts of people to enter the field. Another part of the solution is the need to continually counter the notion that women are innately less capable of mathematical thinking; as long as this attitude persists, many girls will give up before they even give physics a chance. Here the popular media have an important role to play: Film and television producers have the power to create appealing female characters as mathematical brains.

But above all, I suggest that we must also reexamine the long-standing perception of physics as a "transcendent" pursuit for, as I have noted, this view continues to act as a major cultural barrier to women. The solution to this obstacle, I believe, is not to look for ways around it but to remove it altogether. For two and a half thousand years, Western culture has endorsed this view of mathematical science; I suggest it is no longer valid.

The basis for such a view of physics has always been the idea that mathematical relationships discovered in nature are somehow beyond or above nature. Because these relationships have been seen to represent a timeless, immutable core of nature, they have come to be seen as part of a transcendent cosmic plan. Pythagoras saw numbers as the preexisting archetypes for material form, and in the Christian era, scientists from Grosseteste to Newton believed they were illuminating a divine transcendent plan. A similar notion was expressed by Stephen Hawking in the closing pages of *A Brief History of Time*. Like Newton, Hawking suggests the equations of a unified theory exist independently from, and ontologically prior to, the universe itself.

I wish to argue, however, that the mathematical relations we discover in nature are *not* beyond nature, but are merely another *facet* of nature. Rather than having a separate existence apart from or prior to nature, I believe these relations must be viewed like any other scientific discovery. When, for instance, biologists discover the process by which a fertilized egg develops into a child, we do not conclude that the plan for a human child existed in the mind of God before the creation of the universe. When Christian fundamentalists suggest such a notion, physicists are often the first to scoff. Why, then, should we accept that the process by which subatomic particles developed out of a unified force existed in the mind of a deity before the creation of the universe? Why do we give credence to the idea of physics as an illumination of a deity's mind, but not biology? Why, in other words, are we willing to accept that subatomic particles might be preordained by a supernatural power, but not human beings?

I suggest that physics does not warrant the religiously privileged status it has come to have in our culture, and that mathematical knowledge of the world cannot validly be seen as a higher, or transcendent, kind of knowledge, but must be viewed simply as one particular kind of knowledge. By rejecting physicists' claim to a higher form of knowledge, we would, so to speak, bring them down to earth. A by-product of doing so would be to remove one important cultural barrier to women's progress in the field.

Yet my suggestion for "grounding" physics is not driven by a desire for equity, but rather by a growing realization that the transcendent view of physics is no longer tenable. This decision is not one I have come to easily. When I began working on this book several years ago, I was a classic Pythagorean; I too viewed physics as the transcendent science, illuminating a plan that somehow existed beyond nature. It was this feeling that drew me into the field in the first place. But in researching this book and reading extensively in the history and philosophy of science, I have come to see physics in a different light. What I have come to see is that there are many different ways of knowing, and although mathematics is a particularly useful (and to me, still particularly wonderful) way of knowing the world around us, it is *not* more fundamental or "higher" than many other disciplines. In studying physics from the outside, rather than the inside, I have changed my view completely.

The question thus arises: How should we respond to the powerful religious undercurrents in physics today? Again, I suggest that we must reject them.

Before the eighteenth century, mathematical scientists saw themselves as engaged in reading the book of nature, traditionally regarded as God's other book. But none of them lost sight of the fact that first and foremost God was the spiritual Redeemer of humanity, not the Creator of the material world. For them, mathematical science was not a theology on its own, but rather a "handmaiden" to traditional theology—to use Saint Augustine's term. They did not see themselves as a priesthood in their own right. But with the rising power of scientists during the eighteenth century, people began to look to science not so much for support, but for *evidence* of their belief in God. God the Creator began to usurp God the Redeemer. Whereas before people had looked to science merely to *bolster* faith, now belief in God was increasingly grounded in science. Physicists today who equate a theory of everything with "the mind of God" not only ground their "theology" in science, they see themselves as some sort of priesthood *in their own right.* To repeat Einstein's remark quoted earlier: "A contemporary has said, not unjustly, that in this materialistic age of ours the serious scientific workers are the only profoundly religious people."

Jesuit theologian Michael Buckley has pointed out, however, that science cannot form the foundation for a belief in God. When religion turned to science for its foundation, it betrayed itself and, says Buckley, "implicitly confess[ed] its own intrinsic lack of warrant." That remains so today. As Buckley explains: "If religion does not possess the principles and experiences within itself to disclose the existence of God . . . then it is ultimately counterproductive to look outside of the religion to another discipline or science or art to establish that there is a 'friend behind the phenomena.' " Here we have in a nutshell the problem entailed in physicists' claim to be taking us to the mind of God. Those who believe in God cannot look to science for proof of his existence. Religion must find its justification within itself, within what Buckley calls "the phenomenology of religious experience." No set of mathematical equations, no matter how beautiful or embracing, can ever serve as the foundation for a belief in a "friend behind the phenomena." God cannot be found, as Leon Lederman has implied, at the end of an accelerator beam, and those who wish for a genuine rapprochement between the spiritual and the scientific must reject the temptation to accept physics as the basis for their faith.

By the same token, physicists cannot legitimately use religious arguments to justify *their* endeavors. One of the few physicists who has been brave enough to publicly acknowledge this is Steven Weinberg. Just as religion must find its justification within itself, so must science. Science and religion are both valid endeavors but they must not be confused with one another, and we must not allow physicists to bamboozle us with half-baked theologies. To paraphrase Buckley, when physicists turn to religion for their foundation they implicitly confess their own lack of warrant.

Yet Buckley also stresses that to assert that science cannot serve as the foundation for religion is not to imply that it is the enemy of religion. Just as Kepler and Newton saw mathematical relations in nature as fuel for their belief in God, physics today can serve to enhance an already existing belief in a supreme being. While not being able to serve as a *foundation* for faith, physics can still serve as a *handmaiden*. One of the messages I hope to leave in the reader's mind is that one does not have to make a choice between religion and science. Provided one understands the real role of each, and does not let either side take over the other, there is no reason one cannot have both forces in one's life. That is not to imply one must have both, but for the many Americans who do wish for a rapprochement between the two worlds, the message of history is that they are profoundly complementary forces.

Rather than uniting physics with religion, the more difficult and pressing problem I believe we face is how to ground physics in an ethical and socially responsible framework. Traditionally, religion provided this framework, but because the majority of physics practitioners are no longer affiliated with any formal religion, and because a large portion of the research money for physics is now provided by the state (which in this country is formally separated from religion), we need to provide a secular ethical framework for this science. As I suggested earlier, we need a physics that is more centered on human needs and concerns, a physics whose practitioners are more ethically and socially accountable, a physics that is, in effect, more "grounded." Such a suggestion may at first sound radical, but the notion that science should be called to account for itself ethically has become commonplace in the biological sciences; during the last two decades we have seen a proliferation of bioethics centers. If biological science should be subject to ethical scrutiny, why not physics?

Physicists have always claimed that their science is ethically neutral. But in recent years, philosophers of science—particularly feminist philosophers—have challenged this claim. Knowledge, they say, is not neutral, but always the fruit of some intention, whether consciously recognized or not. Recently, Keller has suggested that not only must we reject the myth of ethical neutrality in science, we should choose our scientific projects on the basis of well-thought-out and *conscious* intentions. Rather than leaving physicists to tell us what they want to do and just handing over the money to do it, as a society we must be involved in deciding what we want from physics and what purposes we want it to serve. We must *consciously* move it back to a more socially responsible grounding.

Although, as I have mentioned, it is impossible to predict in advance what difference women would make to the culture and practice of physics, I believe they have an important role to play in helping to make both more ethical. It is not insignificant that it is, in fact, women who are spearheading this call for change. At the beginning of Mathematical Man's story in the time of Pythagoras, he *was* primarily an ethical being. One of the reasons he has lost his ethical grounding is that he has been largely without female company for so long. I do not suggest that a greater presence of women would suddenly turn physics into an ideal science; I only propose that women would provide—as do women in all communities—a balancing influence. As in any society, the best goals emerge from the dreams of men and women together. After two and a half thousand years, the time has come for Mathematical Man to embrace the partnership of Mathematical Woman. The time has come for a mathematically based science envisioned and practiced equally by both sexes.

SOURCES

Introduction

p. 7 – David F. NOBLE, *A World Without Women. The Christian Clerical Culture of Western Science*, NEW YORK: Knopf, 1992, p. 163.

p. 8 – See Stephen William HAWKING, *A Brief History of Time. From the Big Bang to Black Holes*, NEW YORK/LONDON: Bantam Press, 1988.

p. 8 – Paul Charles William DAVIES, *The Mind of God. The Scientific Basis for a Rational World. Science and the Search for Ultimate Meaning.* LONDON/NEW YORK: Simon & Schuster, 1992.

p. 8 – IDEM, *God and the New Physics*, LONDON: Dent, 1983; NEW YORK: Simon & Schuster, 1984.

p. 8 – Ian STEWART, *Does God Play Dice? The Mathematics of Chaos*, OXFORD/CAMBRIDGE (Mass.): Blackwell, 1989, 1990[2].

p. 8 – IDEM/Martin GOLUBITSKY, *Fearful Symmetry. Is God a Geometer?*, OXFORD/CAMBRIDGE (Mass.): Blackwell, 1992.

p. 8 – Leon M. LEDERMAN/Dick TERESI, *The God Particle. If the Universe Is the Answer, What Is the Question?*, BOSTON: Houghton Mifflin, 1993; LONDON: Bantam Press, 1993.

p. 8 – Robert JASTROW, *God and the Astronomers*, NEW YORK/LONDON: Norton, 1978, 1992[2].

p. 10 – Albert EINSTEIN, "Prinzipien der Forschung" (1914); in *Zu Max Plancks 60. Geburtstag. Ansprachen in der Deutschen physikalischen Gesellschaft*, KARLSRUHE: Müller, 1918; later in *Mein Weltbild*, AMSTERDAM: Querido, 1934, p. 169.

p. 11 – See Constance JORDAN, *Renaissance Feminism. Literary Texts and Political Models*, ITHACA (NY): Cornell University Press, 1990.

p. 11 – Sandra HARDING, *The Science Question in Feminism,* ITHACA (NY): Cornell University Press, 1986; MILTON KEYNES: Open University Press, 1986, p. 31.

p. 12 – Albert EINSTEIN, "Religion und Wissenschaft", *New York Times Magazine,* NEW YORK: 9 November 1930, p. 4; also in *Mein Weltbild* cit., p. 42.

p. 14 – Leon M. LEDERMAN/Dick TERESI, *op. cit.,* p. 254.

p. 16 – Elizabeth FEE, "Women's nature and scientific objectivity", in Marian LOWE/Ruth HUBBARD (eds), *Women's Nature. Rationalizations of Inequality,* NEW YORK: Pergamon Press, 1981, p. 22.

Chapter 1. All Is Number

p. 19 – PYTHAGORAS, quoted in ARISTOTLE, *Metaphysica,* I 5. 985b 23–987a 9.

p. 19 – ARISTOTLE, *Perì tôn Pythagoréion* (fr. 192 Rose); quoted in IAMBLICHUS, *De vita Pythagorica liber,* VI, 31; ed. Ludwig DEUBNER, LEIPZIG: Teubner, 1937; new edn, ed. Udalrichus KLEIN, STUTTGART: Teubner, 1975, p. 18.

p. 19 – ARISTOTLE, quoted in *Die Fragmente der Vorsokratiker,* ed. Hermann DIELS/Walther KRANZ, BERLIN: Weidmann, 1903, 1966[12], fr. 14.7, vol. I, pp. 98–99.

p. 20 – Isidore LEVY, *La légende de Pythagore de Grèce en Palestine,* PARIS: Champion, 1927 ("Bibliothèque de L'Ecole des Hautes Études", vol. 250), ch. V ("L'Évangile"), pp. 295–340.

p. 21 – David C. LINDBERG, *The Beginnings of Western Science. The European Scientific Tradition in Philosophical, Religious, and Institutional Context, 600 B.C. to A.D. 1450,* CHICAGO: University of Chicago Press, 1992, p. 13.

p. 21 – IAMBLICHUS, *De vita Pythagorica liber,* III, 13–14; ed. Ludwig DEUBNER/Udalrichus KLEIN cit., pp. 10–11.

p. 21 – PORPHYRIUS, *Vita Pythagorae,* 7; ed. Édouard DES PLACES, PARIS: Les Belles Lettres, 1982, p. 39.

p. 21 – *Ibid,* 8; ed. Édouard DES PLACES cit., p. 39.

p. 22 – *Ibid,* 12; ed. Édouard DES PLACES cit., p. 41.

p. 22 – David C. LINDBERG, *op. cit.*, p. 14.

p. 23 – IAMBLICHUS, *op. cit.*, XVII, 72; ed. Ludwig DEUBNER/ Udalrichus KLEIN cit., p. 41.

p. 23 – PORPHYRIUS *op. cit.*, 18–19; ed. Édouard DES PLACES cit., p. 44.

p. 24 – See Lynn M. OSEN, *Women in Mathematics*, CAMBRIDGE (Mass.): MIT Press, 1974, 1992², pp. 16–17.

p. 24 – See IAMBLICHUS, *op. cit.*, XXXV, 248–251/255–257/ 260 *passim*; ed. Ludwig DEUBNER/Udalrichus KLEIN cit., pp. 133–135/137–138/139–140.

p. 29 – Gerda LERNER, *The Creation of Patriarchy*, NEW YORK: Oxford University Press, 1986, 1987², ch. VII ("The Goddesses"), pp. 141–160.

p. 32 – Albert EINSTEIN, see also note for p.10 above.

p. 35 – Margaret ALIC, *Hypatia's Heritage. A History of Women in Science from Antiquity to the Late Nineteenth Century*, LONDON: The Women's Press, 1986; BOSTON: Beacon Press, 1986, p. 41.

p. 36 – SOCRATES SCHOLASTICUS, *Historia ecclesiastica*, ch. XV ("De Hypatia philosopha"); ed. Henri DE VALOIS, in Jacques-Paul MIGNE (ed.), *Patrologia graeca*, PARIS: 1857–1866, vol. LXVII/ 1864, coll. 769–770.

Chapter 2. God as Mathematician

p. 39 – Suzanne Fonay WEMPLE, *Women in Frankish Society. Marriage and the Cloister 500–900*, PHILADELPHIA: University of Pennsylvania Press, 1981, p. 177.

p. 41 – Walter Jackson ONG, "Latin language study as a Renaissance puberty rite", *Studies in Philology*, CHAPEL HILL (NC): vol. LVI, n. 2/April 1959, p. 107.

p. 42 – *Ibid*, p. 109.

p. 42 – IDEM, *Orality and Literacy. The Technologizing of the Word*, LONDON/NEW YORK: Methuen, 1982, p. 113.

p. 42 – HROTSVITHA, *Legenden*, "Praefatio"; in *Opera*, ed. Helene HOMEYER, PADERBORN: Schöningh, 1970, pp. 38–39.

p. 42 – IDEM, *Dramen,* "Epistola eiusdem ad quosdam sapientes huius libri fautores", in *Opera* cit., pp. 235–236.

p. 43 – HILDEGARD, *Scivias, sive Visionum ac revelationum libri tres* (1141–1151); in Jacques-Paul MIGNE (ed.), *Patrologia latina,* PARIS: 1844–1864, vol. CXCVII/1855 (*Sanctae Hildegardis Abbatissae Opera Omnia*), book I, pref., cols. 383–384; ed. Adelgundis FÜHRKÖTTER/Angela CARLEVARIS, TURNHOUT: Brepols, 1978 ("Corpus Christianorum. Continuatio Mediaevalis", vols. CCCM/XLIII/XLIIIA), "Protestificatio veracium visionum a deo flentium", pp. 3–4.

p. 43 – David F. NOBLE, *op. cit.,* p. 141.

p. 43 – Gerda LERNER, *The Creation of Feminist Consciousness. From the Middle Ages to Eighteen-seventy,* NEW YORK: Oxford University Press, 1993, pp. 55–56

p. 45 – William CLARK, "The misogyny of scholars", *Perspectives on Science. Historical, Philosophical, Social,* CHICAGO: vol. 1, n. 2/1993, p. 349.

p. 45 – Walter MAP [VALERIUS], "Dissuasio Valerii ad Ruffinum philosophum ne uxorem ducat", in *De Nugis Curialium* (ms. 1182), f. 46*v*; ed. Thomas WRIGHT, LONDON: Camden, 1850, p. 150.

p. 45 – Episode quoted in Lynn THORNDIKE, *University Records and Life in the Middle Ages,* NEW YORK: Columbia University Press, 1944, p. 119.

p. 45 – See Londa SCHIEBINGER, *The Mind Has No Sex? Women in the Origins of Modern Science,* CAMBRIDGE (Mass.): Harvard University Press, 1989, ch. III ("Scientific women in the craft tradition"), pp. 66–101.

p. 48 – THIERRY DE CHARTRES, quoted in N. M. HARING, "The creation and creator of the world according to Thierry of Chartres and Clarembaldus of Arras", *Archives d'histoire doctrinale et littéraire du Moyen Age,* PARIS: vol. XXII/1955, p. 196.

p. 48 – Robert GROSSETESTE, quoted in Alistair Cameron CROMBIE, *Robert Grosseteste and the Origins of Experimental Science. 1100–1700,* OXFORD: Clarendon Press, 1953, 1962², p. 102.

p. 48 – IDEM, quoted in Alistair Cameron CROMBIE, *op. cit.,* p. 103.

p. 49 – Matthew PARIS, *Historia maior,* LONDON: Wolf, 1571; ed.

Henry Richard LUARD, LONDON: Longman, 1872–1883, vol. V/1880, p. 227.

p. 50 – DIETRICH VON FREIBERG, *Tractatus de iride et de radialibus impressionibus* (c.1304); ed. Joseph WÜRSCHMIDT, MÜNSTER: Aschendorff, 1914; ed. Maria Rita PAGNONI STURLESE/ Loris STURLESE, in *Opera Omnia*, ed. K. FLASCH/B. MOYSISCH/ R. IMBACH/*et alii*, HAMBURG: Meiner, 1977–1985, vol. IV/ 1985 (*Schriften zur Naturwissenschaft. Briefe*), pp. 95–268.

p. 50 – Roger BACON, *Opus maius ad Clementem Quartum Pontificem Romanum* (1266–1268); ed. John Henry BRIDGES, OXFORD: Clarendon Press, 1897–1900; – *opus minus* (c. 1267); in *Opera quaedam hactenus inedita*, ed. J.S. BREWER, LONDON: Longman/Green/Roberts, 1859, pp. 311–389; – *Opus tertium* (c. 1267); in *Opera quaedam hactenus inedita* cit., pp. 3–310.

p. 52 – IDEM, *Opus maius* cit., vol. I/1897, part IV ("Mathematicae in physicis utilitas"), p. 211.

p. 52 – Samuel Y. EDGERTON Jr., *The Heritage of Giotto's Geometry. Art and Science on the Eve of the Scientific Revolution*, ITHACA (NY): Cornell University Press, 1991, p. 48.

p. 53 – David C. LINDBERG, *op. cit.*, p. 295.

p. 55 – Étienne TEMPIER, quoted in Eduard J. DIJKSTERHUIS, *The Mechanization of the World Picture. Pythagoras to Newton*, PRINCETON (NJ): Princeton University Press, 1986, pp. 173–176.

p. 55 – See David F. NOBLE, *op. cit.*, p. 158.

p. 55 – Christine de PISAN, *Le Livre de la Cité des Dames* (1404–1405); trans. Bryan ANSLAY, LONDON: Pepwell, 1521, ch. I ("Here beginneth the fyrste chapitre whiche telleth howe & by whome the Cyte of Ladies was fyrst begon to buylde"), f. Bbij*v*.

p. 56 – *Ibid*, ch. XXVII ("Cristine demaundeth of reason yf ever god iyste to make a woman so noble to have ony understadynge of the highnesse of science"); trans. cit., f. Kkiiiij*v*.

p. 56 – Eduard J. DIJKSTERHUIS, *op. cit.*, p. 226.

p. 57 – Nicolaus CUSANUS, *De docta ignorantia* (1440); ed. Ernst HOFFMANN/Raymond KLIBANSK, Leipzig: Meiner, 1932; new edn, ed. Paul WILPERT/Hans Gerhard SENGER/Raymond KLIBANSKY, HAMBURG: Meiner, 1964, 1979³, book I, ch. X, p. 21.

p. 57 – IDEM, *Idiota. De sapientia, de mente, de staticis experimentis*, (1450); ed. Ludwig BAUR, in *Opera Omnia*, ed. Ernst HOFFMANN, LEIPZIG: Meiner, 1932–1983, vol. V/1937.

p. 58 – See Shmuel SAMBURSKY, *The Physical World of the Greeks*, trans. Merton DAGUT, LONDON: Routledge & Kegan Paul, 1956; PRINCETON (NJ): Princeton University Press, 1987, p. 223.

p. 58 – Joseph NEEDHAM/Wang LING, *Science and Civilization in China*, CAMBRIDGE: Cambridge University Press, 1956–1984, vol. III/1959 (*Mathematics and the Sciences of the Heavens and the Earth*), sec. 18 ("Mathematics"), K ('Mathematics and science in China and the West'), pp. 150–168.

p. 59 – LEONARDO DA VINCI, ms. An C I 7r/W. 19066; in *Scritti scelti*, ed. Anna Maria BRIZIO, TORINO: UTET, 1952, p. 614.

Chapter 3. Harmony of the Spheres

p. 62 – See Owen GINGERICH, *The Eye of Heaven. Ptolemy, Copernicus, Kepler*, NEW YORK: American Institute of Physics, 1993, p. 163.

p. 62 – Nicolaus COPERNICUS, *De revolutionibus orbium coelestium libri sex*, NÜRNBERG: Petreius, 1543; ed. Ricardus GANSINIEC/ Juliusz DOMANSKI / Jerzy DOBRZYCKI / Alexander BIRKENMAJER, WARSAW/CRACOW: Officina Publica Libris Scientificis, 1975.

p. 63 – Claudius PTOLEMAEUS *Synthaxis mathematica*; ed. J.L. HEIBERG, LEIPZIG: Teubner, 1898.

p. 64 – Nicolaus COPERNICUS, *op. cit.*, pref., f. III*v*; ed. Ricardus GANSINIEC/*et alii* cit., p. 4.

p. 64 – *Ibid*, lib. I. f. lv passim; ed. Ricardus GANSINIEC/*et alii* cit., p. 7.

p. 65 – Fernand HALLYN, *The Poetic Structure of the World. Copernicus and Kepler*, NEW YORK: Zone Books, 1990, p. 94.

p. 65 – Marcus VITRUVIUS Pollio, *De architectura libri decem*, I 2.4; ed. AA. VV, PARIS: Les Belles Lettres, 1969–. . ., vol. I/1990 (ed. Philippe FLEURY), p. 16.

p. 65 – Nicolaus COPERNICUS *op. cit.*, pref., f. III*v*; ed. Ricardus GANSINIEC/*et alii* cit., p. 4.

p. 67 – Tycho BRAHE, letter to Christopher ROTHMANN, August 1590; in *Epistularum astronomicarum libri*, URANIBORG: 1596, p. 191; ed. G. A. HAGEMANN/Johann RAEDER, in *Opera Omnia*, ed. John Louis Emil DREYER, KOBENHAVN: Gyldendal, 1913–1929, vol. VI, t. I/1919, pp. 221–222.

p. 67 – Owen GINGERICH, *op. cit.*, p. 183.

p. 68 – Fernand HALLYN *op. cit.*, p. 136.

p. 69 – Isaac NEWTON, letter to Robert HOOKE, 5 February 1675 [1676]; in *The Correspondence*, ed. H. W. TURNBULL/*et alii*, CAMBRIDGE: Cambridge University Press, 1959–1977, vol. I/ 1959 (*1661–1675*), letter 154, p. 416.

p. 69 – Johannes KEPLER, *Prodromus dissertationum mathematicarum continens Mysterium cosmographicum, de admirabili proportione orbrium coelestium*, TÜBINGEN: Gruppenbach, 1596, pref., p. 6; ed. Max CASPAR, in *Gesammelte Werke*, Walther von DYCK/Max CASPAR, MÜNCHEN: Beck, 1937–. . ., vol. I/1938, p. 9.

p. 70 – Arthur KOESTLER, *The Sleepwalkers. A History of Man's Changing Vision of the Universe*, LONDON: Hutchinson, 1959; NEW YORK: Macmillan, 1959, p. 234.

p. 71 – Johannes KEPLER, *Harmonice mundi libri quinque*, FRANKFURT AM MAIN/LINZ: Tampach, 1619, book IV, ch. I, p. 119; ed. Max CASPAR, in *Gesammelte Werke* cit., vol. VI/1940, p. 223.

p. 71 – IDEM, *De fundamentis astrologiae certioribus*, PRAHA: Schuman, 1602, f. Bij*v*; ed. Max CASPAR/Franz HAMMER, in *Gesammelte Werke* cit., vol. IV/1941 (*Kleinere Schriften 1602–1611. Dioptrice*), p. 15.

p. 71 – IDEM, *Mysterium cosmographicum* cit., ch. I, p. 19; ed. Max CASPAR, in *Gesammelte Werke* cit., vol. I/1938, p. 23.

p. 71 – IDEM, *Harmonice mundi* cit., book IV, ch. I, p. 119; in *Gesammelte Werke* cit., vol. VI/1940, p. 223.

p. 71 – IDEM, *Astronomiae pars optica*, FRANKFURT AM MAIN: Marnius/Auber, 1604, pref., f. 3; ed. Franz HAMMER, in *Gesammelte Werke* cit., vol. II/1939, p. 16.

p. 71 – IDEM, letter to Michael MÄSTLIN, 3 October 1595; ed. Max CASPAR, in *Gesammelte Werke* cit., vol. XIII/1945 (*Briefe 1590–1599*), letter 23, p. 40.

p. 74 – IDEM, *Astronomia nova aitiologétos, seu Physica coelestis, tradita commentariis de motibus stellae Martis*, PRAHA: s.e., 1609; HEIDELBERG: Vögelin, 1609, part IV, ch. 58, p. 285; ed. Max CASPAR, in *Gesammelte Werke* cit., vol III/1937, p. 366.

p. 77 – IDEM, *Tabulae Rudolphinae*, ULM: Saur, 1627; ed. Franz HAMMER, in *Gesammelte Werke* cit., vol. 10/1696.

p. 77 – Elias von LÖWEN, "Maritus ad lectorem", in Maria CUNITZ, *Urania propitia, sive Tabulae astronomicae mire faciles, wim hypothesium physicarum a Kepplero proditarum complexae*, OELS: Seyffert, 1650, ff. [11 *v*-12*r*].

p. 77 – Elisabetha KOOPMAN, "Epistola dedicatoria" in Johannes HEVELIUS, *Prodromus astronomiae, exhibens fundamenta, quibus additus est uterque catalogus stellarum fixarum*, DANZIG: Stoll, 1690, ff. 1 *r*-3*v*.

p. 78 – Gottfried KIRCH, *Kurtze Betrachtung derer Wunder am gestirnten Himmel, welche veranlasset der itzige recht merkwürdige neue Comet*, LEIPZIG: Kirchner, 1677; *Neue Himmels-Zeitung darinnen von den zweien neuen großen im 1680. Jahr erschienenen Cometen*, NÜRNBERG: Endters, 1681.

p. 79 – Johann T. JABLONSKI, letter to Gottfried Wilhelm LEIBNIZ, 1 November 1710; in Adolf HARNACK, "Berichte des Secretars der brandenburgischen Societät der Wissenschaften Johann T. Jablonski an den Präsidenten Gottfried Wilhelm Leibniz (1700–1715) nebst einigen Antworten von Leibniz", *Philosophische und historische Abhandlungen der königlichen Akademie der Wissenschaften*, BERLIN: vol. III/1897, pp. 79–80.

Chapter 4. The Triumph of Mechanism

p. 81 – Nicolaus COPERNICUS, *op. cit.*, pref., f. IV*r*/book. I, f. 10*r*, ed. Ricardus GANSINIEC/*et alii* cit., pp. 4–5/20.

p. 83 – *Corpus Hermeticum*, ed. Arthur Darby NOCK/André Marie Jean FESTUGIERE, PARIS: Les Belles Lettres, 1946–1954.

p. 84 – See Frances Amelia YATES, *Giordano Bruno and the Hermetic Tradition*, LONDON: Routledge & Kegan Paul, 1964; CHICAGO: University of Chicago Press, 1979, p. 6.

p. 85 – IDEM, *op. cit.*, pp. 2 et seq.

p. 86 – *Picatrix* (XII sec.); Latin trans. XV cent., book II, ch. 10; quoted in Frances Amelia Yates, *op. cit.*, pp. 52–53.

p. 86 – Marsilius FICINUS, *De vita libri tres*, FIRENZE: Mischomino, 1489.

p. 88 – Keith THOMAS, *Religion and the Decline of Magic. Studies in Popular Beliefs in Sixteenth and Seventeenth Century England*, LONDON: Weidenfeld & Nicolson, 1971; NEW YORK: Scribner, 1971, p. 320.

p. 88 – Walter RALEIGH, *The History of the World*, LONDON: Burre, 1614, ch. XI ("Of Zoroaster, supposed to have beene the chiefe author of magick arts: and of the divers kinds of magicke"), II ("Of the name of magia: and that it was anciently farre divers from coniuring and witchcraft"), p. 201.

p. 88 – Petrus GARSIAS, *In determinationes magistrales contra conclusiones apologales Joannes Pici Mirandulani concordie comitis*, ROMA: Silber, 1489.

p. 90 – David F. NOBLE, *op. cit.*, p. 197.

p. 90 – Carolyn MERCHANT, *The Death of Nature. Women, Ecology and the Scientific Revolution*, NEW YORK: Harper & Row, 1980; LONDON: Wildwood House, 1982, ch. 5.2 ("Disorder, sexuality, and the witch"), pp. 132–140.

p. 91 – Frances Amelia YATES, *op. cit.*, pp. 444–445.

p. 91 – David F. NOBLE, *op. cit.*, p. 218.

p. 93 – Pierre GASSEND, *Epistolica exercitatio, in qua Principia philosophiae Robert Fluddi Medici reteguntur; et ad recentes illius Libros, adversus R. P. F. Marinum Mersennum Ordinis Minimorum Sancti Francisci de Paula scriptos respondetur,*

PARIS: Cramoisy, 1630; in *Opera Omnia*, ed. Henri Ludovic HABERT DE MONTMOR/F. HENRI, LYON: Anisson/Devenet, 1658, vol. III (*Philosophica Opuscula*), pp. 211–268.

p. 93 – Peter DEAR, *Mersenne and the Learning of the Schools*, ITHACA (NY): Cornell University Press, 1988, pp. 3–4.

p. 93 – Pierre GASSEND, *Syntagma Philosophicum* (1658); in *Opera Omnia* cit., vol. I, p. 158.

p. 93–4 – William B. ASHWORTH Jr., "Catholicism and early modern science", in DAVID C. LINDBERG/Ronald L. NUMBERS (eds), *God and Nature. Historical Essays on the Encounter between Christianity and Science*, BERKELEY: University of California Press, 1986, p. 141.

p. 94 – Carolyn MERCHANT, *op. cit.*, ch. 5 ("Nature as disorder. Women and witches"), pp. 127–148.

p. 94 – René DESCARTES, letter to Marin MERSENNE, 15 April 1630; in *Oeuvres* cit., vol. I/1897 (*Correspondence: 1 Avril 1622–Fé vrier 1638*), letter XXI, p. 135; in *Correspondance*, ed. Charles ADAM/Gérard MILHAUD, PARIS: Alcan/Presses Universitaires de France, 1936–1963, vol. I/1936, letter 28, p. 129.

p. 94 – Carolyn MERCHANT, *op. cit.*, p. 205.

p. 94 – Thomas HOBBES, *Leviathan, or the Matter, Forme and Power of a Commonwealth, Ecclesiasticall and Civill*, LONDON: Crooke, 1651; ed. Richard TUCK, CAMBRIDGE: Cambridge University Press, 1991.

p. 94 – Carolyn MERCHANT, *op. cit.*, p. 209.

p. 95 – Edwin Arthur BURTT, *The Metaphysical Foundations of Modern Physical Science. A Historical and Critical Essay*, LONDON: Kegan Paul, 1924, LONDON: Routledge & Kegan Paul, 1932²; NEW YORK: Harcourt Brace, 1925, 1932²; NEW YORK: Humanities Press, 1951, p. 105.

p. 96 – Adrien BAILLET, *La vie de Monsieur Descartes*, PARIS: Horthemels, 1691, vol. I, pp. 81/115.

p. 96 – René DESCARTES, *Discours de la méthode, pour bien conduire sa raison, et chercher la vérité dans les sciences*, LEIDEN: Maire, 1637, part IV, p. 33 [Latin trans. ed. Éstienne de COURCELLE, *Specimina philosophiae, seu Dissertatio de methodo recte regendae rationis et veritatis in scientia investigandae*, AMSTERDAM:

Elzevier, 1644]; ed. Étienne GILSON, PARIS: Vrin, 1925, 1967⁴, p. 32.

p. 96 – Frances Amelia YATES, *op. cit.*, pp. 454–455.

p. 98 – Marguerite YOURCENAR.

p. 98 – David F. NOBLE, *op. cit.*, pp. 212–215.

p. 99 – Martha ORNSTEIN, *The Role of Scientific Societies in the Seventeenth Century*, CHICAGO: University of Chicago Press, 1928, p. 75.

p. 99 – Federico CESI, *Lynceographum* (1605); quoted in Domenico CARUTTI, *Breve storia della Accademia dei Lincei*, ROMA: Salviucci, 1883, p. 7.

p. 99 – John EVELYN, letter to Robert BOYLE, 3 September 1659; in *The Works*, ed. Thomas BIRCH, LONDON: Millar, 1744, vol. V, p. 398; also in *Diary and Correspondence*, ed. William BRAY, LONDON/NEW YORK: Routledge/Dutton, 1906, p. 590.

p. 100 – David F. NOBLE, *op. cit.*, p. 225.

p. 100 – Walter CHARLETON, *The Ephesian Matron*, LONDON: Herringman, 1659, p. 112.

p. 100 – Henry (Heinrich) OLDENBURG, in Robert BOYLE, *Experiments and Considerations Touching Colours*, LONDON: Herringman, 1664, "The Publisher to the Reader", f. A8*r*.

p. 100 – Londa SCHIEBINGER, *op. cit.*, p. 26.

p. 101 – René DESCARTES, *Principia philosophiae*, AMSTERDAM: Elzevier, 1644, "Serenissimae Principi Elisabethae", f. *3*v.

p. 102 – Londa SCHIEBINGER, *op. cit.*, p. 170.

p. 102 – François POULAIN de la BARRE, *De l'égalité de deux sexes. Discours physique et morale, ou l'on voit l'importance de se défaire des préjugez*, PARIS: Du Puis, 1673, part I ("Où l'on montre que l'opinion vulgaire est un préjugé, & qu'en comparant sans interest ce que l'on peut remarquer dans la conduite des hommes & des femmes, on est obligé de reconnoître entre les deux sexes une égalité entiere"), pp. 1–75.

p. 102-3 – Margaret LUCAS CAVENDISH, *Poems and Fancies*, LONDON: Martin/Allestrye, 1653;

– *Philosophical Fancies*, LONDON: Martin/Allestrye, 1653;

- *The Philosophical and Physical Opinions*, LONDON: Martin/ Allestrye, 1655; LONDON: Wilson, 1663; also *Grounds of Natural Philosophy*, LONDON: Maxwell, 1668;
- *Nature's Pictures Drawn by Fancies Pencil to the Life*, LONDON: Martin/Allestrye, 1656; LONDON: Maxwell, 1671;
- *Philosophical Letters, or Modest Reflections upon Some Opinions in Natural Philosophy, Maintained by Several Famous and Learned Authors of This Age*, LONDON: s.e., 1664;
- *Observations upon Experimental Philosophy, to which is added, the Description of a New Blazing World*, LONDON: Maxwell, 1666; LONDON: Maxwell, 1668.

Chapter 5. The Ascent of Mathematical Man

p. 105 – René DESCARTES, *Meditationes de prima philosophia, in qua Dei existentia et animae immortalitas demonstrantur*, PARIS: Soly: 1641; also *Meditationes de prima philosophia, in quibus Dei existentia, et animae humanae a corpore distinctio, demonstrantur*, AMSTERDAM: Elzevier, 1642.

p. 107 – Arthur KOESTLER, *op. cit.*, p. 354.

p. 109 – Galileo GALILEI, *Sidereus Nuncius*, VENEZIA: Baglioni, 1610; in *Opere*, ed. Franz BRUNETTI, TORINO: UTET, 1964, 1980², vol. I, pp. 263–319.

p. 109 – Quoted in Arthur KOESTLER, p. 369.

p. 110 – See Mario BIAGIOLI, *Galileo Courtier. The Practice of Science in the Culture of Absolutism*, CHICAGO: University of Chicago Press, 1993.

p. 111 – Galileo GALILEI, *Dialoghi sopra i due massimi sistemi del mondo, tolemaico e copernicano*, FIRENZE: Landini, 1632; in *Opere* cit., vol. II, pp. 7–552.

p. 111 – IDEM, *Discorsi e dimostrazioni matematiche intorno a due nuove scienze attenenti alla mecanica et i movimenti locali*, LEIDEN: Elzevier, 1638; ed. Enrico GIUSTI, TORINO: Einaudi, 1990.

p. 113 – Edwin Arthur BURTT, *op. cit.*, p. 38.

p. 115 – Richard Samuel WESTFALL, *Never at Rest. A Biography of Isaac Newton,* CAMBRIDGE: Cambridge University Press, 1980, p. 58.

p. 115 – *Ibid,* p. 62.

p. 117 – *Ibid,* p. 155.

p. 120 – Isaac NEWTON, *Philosophiae naturalis principia mathematica,* LONDON: Streater, 1687; CAMBRIDGE: Cotes, 1713; LONDON: Innys, 1726; ed. Alexandre KOYRE/I. Bernard COHEN, CAMBRIDGE: Cambridge University Press, 1972.

p. 120 – Derek GJERTSEN, "Newton's success", in John FAUVEL/ Raymond FLOOD/Michael SHORTLAND/Robin WILSON (eds), *Let Newton Be! A New Perspective on His Life and Works,* OXFORD: Oxford University Press, 1988, p. 35.

p. 120 – Charles-Louis de MONTESQUIEU, *L'esprit des lois,* GENEVE: Barrillot, 1748.

p. 120 – Isaac NEWTON, *Theological Manuscripts,* ed. Herbert McLACHLAN, LIVERPOOL: Liverpool University Press, 1950.

p. 121 – IDEM, letter to Richard BENTLEY, 10 December 1692; in *The Correspondence* cit., vol. III/1961 (*1688–1694*), letter 398, p. 233.

p. 122 – IDEM, *Philosophiae naturalis principia mathematica* cit., "Scholium generale", p. 527; ed. Alexandre KOYRÉ/I. Bernard COHEN cit., vol. II, p. 760.

p. 123 – *Ibid,* p. 529; Alexandre KOYRÉ/I. Bernard COHEN cit., vol. cit., p. 763.

p. 123 – *Ibid,* p. 528; ed. Alexandre KOYRÉ/I. Bernard COHEN cit., vol. cit., p. 761.

p. 124 – Edwin Arthur BURTT, *op. cit.,* p. 284.

p. 124 – Penelope GOUK, "The harmonic roots of Newtonian science", in John FAUVEL/Raymond FLOOD/Michael SHORTLAND/ Robin WILSON (eds), *Let Newton Be!* cit., p. 120.

p. 124 – Isaac NEWTON, quoted in David GREGORY, *Memoranda*; in *The Correspondence of Isaac Newton* cit., vol. III cit., 5/6/ 7 March 1694, pp. 336–338.

p. 125 – Piyo RATTANSI, "Newton and the wisdom of the ancients", in John FAUVEL/Raymond FLOOD/Michael SHORTLAND/ Robin WILSON (eds), *Let Newton Be!* cit., p. 199.

p. 126 – Richard Samuel WESTFALL, *op. cit.,* pp. 407–408.

Chapter 6. God, Women, and the New Physics

p. 128 – John Craig, *Theologiae Christianae principia mathematica*, London: Child, 1699.

p. 128 – *Ibid*, p. 10.

p. 128 – Derek Gjertsen, *op. cit.*, in John Fauvel/Raymond Flood / Michael Shortland / Robin Wilson (eds), *Let Newton Be!* cit., p. 31.

p. 129 – *Ibid*, loc. cit.

p. 129 – Margaret C. Jacob, "Christianity and the Newtonian worldview", in David C. Lindberg/Ronald L. Numbers (eds), *God and Nature* cit., p. 243.

p. 129–30 – Samuel Clarke, *Sixteen Sermons on the Being and Attributes of God, the Obligations of Natural Religion, and the Truth and Certainty of the Christian Revelation, preached in the Years 1704 and 1705, at the Lecture founded by the Honourable Robert Boyle*, ed. John Clarke, in *The Works*, London: Knapton, 1738, vol. II, pref., p. 517.

p. 130 – Roger Hahn, "Laplace and the mechanistic universe", in David C. Lindberg/Ronald L. Numbers (eds), *God and Nature* cit., p. 263.

p. 130 – Bernard Nieuwentijt, *Het Regt Gebruik der Werelt Beschouwingen*, Amsterdam: Wolters & Pauli, 1717.

p. 132 – Margaret C. Jacob, *op. cit.*, in David C. Lindberg/ Ronald L. Numbers (eds), *God and Nature* cit., p. 245.

p. 132–3 – *Ibid.*, loc. cit.

p. 134 – Mary Terrall, "Gendered spaces, gendered audiences. Inside and outside the Paris Academy of Sciences", lecture given to the Clark Library Workshop on *Gender and Science in Early Modern Europe*, Los Angeles: February 1994.

p. 134 – Bernard Le Bovier de Fontenelle, *Entretiens sur la pluralité des mondes*, Paris: Blageart, 1686; ed. A. Calame, Paris: Didier, 1966; Paris: Stfm, 1991.

p. 134 – Mary Terrall, *op. cit.*

p. 134 – Francesco Algarotti, *Il neutonianismo per le dame, ovvero Dialoghi sopra la luce in colori*, Napoli: Pasquali, 1737 (later *Dialoghi sopra l'ottica neutoniana*); ed. Ettore Bonora, in

Opere, MILANO/NAPOLI, Ricciardi, vol. XLVI/1970 (*Illuministi italiani*, vol. II. *Opere di Francesco Algarotti e di Saverio Betttinelli*), pp. 11–177; also TORINO: Einaudi, 1977.

p. 135 – Mary TERRALL, *op. cit.*

p. 135 – François-Marie AROUET de VOLTAIRE, *Eléments de la philosophie de Neuton. Mis à la portée de tout le monde*, AMSTERDAM: Ledet, 1738.

p. 135 – Gabrielle-Émilie LE TONNELIER DE BRETEUIL du CHATELET, *Institutions de physique*, PARIS: Prault, 1740.

p. 136 – Isaac NEWTON, *Philosophiae naturalis principia mathematica*, French trans. Gabrielle-Émilie LE TONNELIER DE BRETEUIL du CHATELET, *Principes mathématiques de la philosophie naturelle*, PARIS: Desaint & Saillant, 1759.

p. 136 – René DESCARTES, *Principia philosophiae*.

p. 136 – *Ibid*, "La traduttrice ai lettori", p. [VIII].

p. 137 – Stephen HALES, *Vegetable Staticks, or An Account of Some Statical Experiments on the sap in Vegetables. Also a Specimen of an Attempt to Analyse the Air*, LONDON: Innys, 1727.

p. 137 – Paula FINDLEN, "Translating the new science. Women and the circulation of knowledge in Enlightenment Italy", lecture given to the Clark Library Workshop on *Gender and Science in Early Modern Europe* cit.

p. 137 – Quoted in Paula FINDLEN, "Science as a cancer in Enlightenment Italy. The strategies of Laura Bassi", *Isis. An International Review devoted to the History of Science and its Cultural Influences*, CHICAGO: vol. 84, no. 3/September 1993, p. 448.

p. 138 – EADEM, "Science as a cancer in Enlightenment Italy. The strategies of Laura Bassi", *Isis. An International Review devoted to the History of Science and its cultural Influences*, CHICAGO: vol. 84, n. 3/September 1993, pp. 450–451.

p. 138–9 – *Ibid*, pp. 454, 464, 468.

p. 140–1 – *Ibid*, p. 467.

p. 141 – Maria Gaetana AGNESI, *Propositiones philosophicae*, MILANO: Malatesta, 1738; – *Instituzioni analitiche ad uso della gioventù italiana*, MILANO: Regia Ducal Corte, 1748.

p. 142 – Pierre-Simon de LAPLACE, *Théorie de Jupiter et de Saturne*, PARIS: Imprimerie Royale, 1787.

p. 143 – Immanuel KANT, *Allgemeine Naturgeschichte und Theorie des Himmels*, KÖNIGSBERG/LEIPZIG: Petersen, 1755; in *Werke*, ed. Wilhelm WEISCHEDEL, WIESBADEN: Insel, 1956–1964, vol. I/1960, pp. 219–396.

p. 143 – Pierre-Simon de LAPLACE, *Exposition du système du monde*, PARIS: Cercle Social, 1796.

p. 144 – Pierre-Simon de LAPLACE, quoted in Roger HAHN, "Laplace and the mechanistic universe", in David C. LINDBERG/Ronald L. NUMBERS (eds), *God and Nature* cit., p. 276.

p. 144 – IDEM, *Mémoires de mathématique et de physique presentés à l'Académie Royale des Sciences par divers savants*, PARIS: Imprimerie Royale, a. VII/1773, p. 113.

p. 144 – Roger HAHN, *op. cit.*, in David C. LINDBERG/Ronald L. NUMBERS (eds) *God and Nature* cit., p. 269.

p. 145 – Edmund HALLEY, "In viri praestantissimi Isaaci Newtoni", in Isaac NEWTON, *Philosophiae naturalis principia mathematica* cit.; ed. Alexandre KOYRÉ/I. Bernard COHEN cit, p. 14.

p. 146 – Immanuel KANT, *Beobachtungen über das Gefühl des Schönen und Erhabenen*, KÖNIGSBERG: Kanter, 1764, pp. 51–2 *passim*; in *Werke* cit., vol. cit., p. 852.

p. 146 – Christoph MEINERS, *Geschichte des weiblichen Geschlechts*, HANNOVER: Helwing, 1788–1800.

p. 146 – Jean-Jacques ROUSSEAU, *A M. d'Alembert sur son article "Genève", dans le VII vol. de l'"Encyclopédie", et particulièrement sur le project d'établir un théâtre de comédie en cette ville (20 mars 1758)*, Amsterdam: Rey, 1758; ed. Bernard GAGNEBIN/Jean ROUSSET, *J.-J. Rousseau citoyen de Genève à M. d'Alembert*, in *Oeuvres complètes*, ed. Bernard GAGNEBIN/Marcel RAYMOND, Paris: Gallimard, 1959–1995, vol. V/1995 (*Ecrits sur la musique, la langue et le théâtre*), p. 96.

p. 147 – Denis DIDEROT, *Sur les femmes* (1772); in *Oeuvres complètes*, ed. J. ASSEZAT, PARIS: Garnier, 1875–1877, vol. II/1875, p. 262.

p. 147 – Georges-Louis Leclerc de BUFFON, *Histoire naturelle, générale et particulière*, PARIS: Imprimerie Royale, 1749–1804.

p. 147 – Londa SCHIEBINGER, *op. cit.*, pp. 153, 154.

p. 148 – *Ibid*, p. 236.

p. 149 – Joan B. LANDES, *Women and the Public Sphere in the Age of the French Revolution*, ITHACA (NY): Cornell University Press, 1988, ch. 4 ("Women and the Revolution") pp. 93–151.

Chapter 7. Science as Salvation

p. 153 – Francis BACON, *Essayes. Religious Meditations. Places of Perswasion and Disswasion*, LONDON: Hooper, 1597; also *Essayes*, LONDON: Beale, 1612; also *The Essayes or Counsels Civill and Morall*, LONDON: Barrett(/Whitaker), 1625; ed. Mario MELCHIONDA, *Gli "Essayes" di Francis Bacon. Studio introduttivo, testo critico e commento*, FIRENZE: Olschki, 1979.

p. 154 – IDEM, *Valerius Terminus. Of the Interpretation of Nature* (ms. 1603); in *Letters and Remains of the Lord Chancellor*, ed. Robert STEPHENS, LONDON: Bowyer, 1734, ch. I ("Of the ends and limits of knowledge"), pp. 406–407.

p. 154 – IDEM, *Temporis partus masculus*, (ms. 1602–1603); in *Scripta in naturali et universali philosophia*, ed. Isaac GRUTER, AMSTERDAM: Elzevier, 1653.

p. 154 – IDEM, *New Atlantis. A Worke Unfinished* (ms. 1614–1617); ed. William RAWLEY in *Sylva Sylvarum. Or a Naturall Historie in Ten Centuries*, LONDON: Lee, 1627; ed. Alfred B. GOUGH, OXFORD: Clarendon Press, 1915.

p. 154–5 – *Ibid*, p. 26; ed. Alfred B. GOUGH cit., p. 32.

p. 155 – *Ibid*, pp. 22–24 *passim*, 31; ed. Alfred B. GOUGH cit., pp. 27–29, 35.

p. 156 – *Ibid*, pp. 35–36 *passim*; ed. Alfred B. GOUGH cit., p. 38–39.

p. 156 – *Ibid*, pp. 31–32; ed. Alfred B. GOUGH cit., p. 36.

p. 156 – *Ibid*, p. 45; ed. Alfred B. GOUGH cit., p. 46.

p. 156 – *Ibid*, p. 29 *passim*; ed. Alfred B. GOUGH cit., p. 33.

p. 159 – Albert EINSTEIN, *Autobiographisches – Autobiographical Notes*, in *The Library of Living Philosophers*, ed. Paul Arthur

SCHILPP, EVANSTON (Ill.): s.e., 1939–1967, vol. VII/1949 (*Albert Einstein: Philosopher-Scientist*), pp. 32–35.

p. 161 – Alfred Russel WALLACE, *The Wonderful Century. Its Successes and Its Failures*, LONDON: Swan Sonnenschein, 1898; NEW YORK: Dodd/Mead, 1898.

p. 161 – *Ibid*, pref., p. VII.

p. 161 – John Burdon Sanderson HALDANE, *Possible Worlds and Other Papers*, LONDON: Chatto and Windus, 1927; NEW YORK: Harper, 1928, p. 302.

p. 162 – Ann BRAUDE, *Radical Spirits. Spiritualism and Women's Rights in Nineteenth-Century America*, BOSTON: Beacon Press, 1989, pp. 4–5.

p. 162 – Charles Robert DARWIN, *On the Origin of Species by Means of Natural Selection, or the Preservation of Favoured Races in the Struggle for Life*, LONDON: Murray, 1859.

p. 162 – John William DRAPER, *History of the Conflict between Religion and Science*, NEW YORK: Appleton, 1872; LONDON: King, 1875, pp. x–xi, 335/364.

p. 163 – Andrew Dickson WHITE, *A History of the Warfare of Science with Theology in Christendom*, NEW YORK: Appleton 1896; LONDON: Macmillan, 1896, p. 7.

p. 163 – See Jeffrey Burton RUSSELL, *Inventing the Flat Earth. Columbus and Modern Historians*, WESTPORT (Conn.): Praeger, 1991, pp. xiii/2.

p. 164 – John William DRAPER, *op. cit.*

p. 164 – See Mary MIDGLEY, *Science as Salvation. A Modern Myth and Its Meaning*, LONDON: Routledge, 1992.

p. 166 – Pierre-Simon de LAPLACE, *Mécanique céleste*, trans. Mary FAIRFAX SOMERVILLE, *Mechanism of the Heavens*, LONDON: Murray, 1831.

p. 166 – Margaret Walsh ROSSITER, *Women Scientists in America. Struggles and Strategies to 1940*, BALTIMORE: Johns Hopkins University Press, 1982, p. 9.

p. 167 – *Ibid*, p. xvi.

p. 167 – Harriet BROOKS, letter to the Principal, Laura GILL, 18 July 1906; in *Departmental Correspondence 1906–1908*, file 41 (Barnard College Archives).

p. 168 – Laura GILL, letter to Harriet Brooks, 23 July 1906; in *Departmental Correspondence 1906–1908* cit., loc. cit.

p. 169 – Mary W. WHITNEY, "Scientific study and work for women", *Education*, MOBILE (AL): a. III/1882, p. 67.

p. 171 – Helena M. PYCIOR, "Marie Curie's 'anti-natural path': time only for science and family", in Pnina G. ABIR-AM/Dorinda OUTRAM (eds), *Uneasy Careers and Intimate Lives. Women in Science 1789–1978*, NEW BRUNSWICK (NJ): Rutgers University Press, 1987, p. 199.

p. 171 – Eve CURIE, *Madame Curie*, PARIS: Gallimard, 1938, p. 125.

Chapter 8. The Saint Scientific

p. 176 – *The Washington Post*, WASHINGTON: 18 April 1955.

p. 177 – Albert EINSTEIN, *Autobiographisches* cit., pp. 8–9.

p. 178 – IDEM, "Folgerungen aus den Kapillaritätserscheinungen", *Annalen der Physik*, LEIPZIG: ser. 4, vol. IV/1901, pp. 513–523;
– "Thermodynamische Theorie der Potentialdifferenz zwischen Metallen und vollständig dissoziierten Lösungen ihrer Salze, und eine elektrische Metode zur Erforschung der Molekularkräfte", *Annalen der Physik*, LEIPZIG: ser. cit., vol. VIII/1902, pp. 798–814;
– "Kinetische Theorie des Wärmegleichgewichtes und des zweiten Hauptsatzes der Thermodynamik", *Annalen der Physik*, LEIPZIG: ser. cit., vol. IX/1902, pp. 417–433.

p. 178 – IDEM, letter to Michele BESSO, 12 December 1919; in IDEM/ Michele BESSO, *Correspondance 1903–1955*, ed. Pierre SPEZIALI, PARIS: Hermann, 1972, letter 51 (E. 41), pp. 147–149.

p. 182 – IDEM, letter to Arnold SOMMERFIELD, 29 October 1912; in IDEM/Arnold SOMMERFIELD, *Briefwechsel. Sechzig Briefe aus dem goldenen Zeitalterer der modernen Physik*, ed. Arnim HERMANN, BASEL: Schwabe, 1968, letter [1], p. 26.

p. 185 – PLATO, quoted in PLUTARCH, *Quaestiones conviviales*, VIII 2; ed. C. HUBERT, in *Moralia*, ed. AA. Vv, LEIPZIG: Teubner, 1938, vol. IV, p. 261.

p. 185 – Albert EINSTEIN, quoted in Ilse ROSENTHAL-SCHNEIDER, *Reality and Scientific Truth*, DETROIT: Wayne State University Press, 1980, p. 74.

p. 185 – IDEM, quoted in Abraham PAIS, *"Subtle Is The Lord . . ."* *The Science and the Life of Albert Einstein*, OXFORD: Clarendon Press, 1982, pp. 30/113.

p. 185 – IDEM, letter to Max BORN, 4 December 1926; in IDEM/ Hedwig BORN/Max BORN, *Briefwechsel 1916–1955*, MÜNCHEN: Nymphenburger, 1969, letter no. 52, pp. 129–130.

p. 186 – IDEM, "Science and Religion", *Nature*, LONDON: vol. 146, no. 605/1940 or 41?, p. 605; also in *Out of My Later Years*, LONDON: Thames and Hudson, 1950; NEW YORK: Philosophical Library, 1950, p. 26.

p. 186 – IDEM, "Religion und Wissenschaft" cit., p. 3; also in *Mein Weltbild* cit., p. 40.

p. 187 – IDEM, "Prinzipien der Forschung" cit., in *Zu Max Plancks 60. Geburtstag. Ansprachen in der Deutschen physikalischen Gesellschaft* cit.; also in *Mein Weltbild* cit., pp. 168–169 *passim*.

p. 187 – Abraham PAIS, *"Subtle Is the Lord . . ."* cit.

p. 187 – Banesh HOFFMANN/Helen DUKAS, *Albert Einstein. Creator and Rebel*, NEW YORK: Viking, 1972; LONDON: Hart-Davis, MacGibbon, 1973.

p. 187 – Carl SEELIG, *Albert Einstein. Eine dokumentarische Biographie*, ZÜRICH: Europa, 1960.

p. 187 – Albert EINSTEIN, quoted in Hedwig BORN, letter to Albert EINSTEIN, 9 October 1944; in IDEM/Hedwig BORN/Max BORN, *Briefwechsel 1916–1955* cit., letter no. 82, p. 209.

p. 188 – IDEM, quoted in Alexander MOSZKOWSKI, *Einstein. Einblicke in seine Gedankenwelt*, HAMBURG/BERLIN: Hoffmann und Campe/Fontane, 1921, p. 87.

p. 190 – Sharon BERTSCH MCGRAYNE, *Nobel Prize Women in Science. Their Lives, Struggles, and Momentous Discoveries*, NEW YORK: Birch Lane Press, 1993, p. 64.

p. 191 – Quoted in Sharon BERTSCH MCGRAYNE, *op. cit.*, p. 68.

p. 191 – Hermann WEYL, "Obituary of Emmy Noether", in Auguste DICK, *Emmy Noether, 1882–1935*, BASEL: Birkhäuser, 1970.

p. 192 – Episode quoted in Sharon BERTSCH MCGRAYNE, *op. cit.*, p. 72.

p. 193 – Hermann WEYL, *op. cit.*, p. 112.

p. 194 – See Sharon BERTSCH MCGRAYNE, *op. cit.*, p. 38.

p. 194 – Max PLANCK, quoted in Sharon BERTSCH MCGRAYNE, *op. cit.*, p. 43.

p. 194 – Episode quoted in Sharon BERTSCH MCGRAYNE, *op. cit.*, pp. 42–43.

p. 195 – Albert EINSTEIN, quoted in Sharon BERTSCH MCGRAYNE, *op. cit.*, p. 48.

p. 196 – Lise MEITNER/Otto Robert FRISCH, "On the products of the fission of uranium and thorium under neutron bombardment", *Det Kgl. Danske Videnskabernes Selskab. Mathematisk-fysiske Meddelelser.*, vol. 17, no. 5/1939.

Chapter 9. Quantum Mechanics and a "Theory of Everything"

p. 201 – Albert EINSTEIN, "Einheitliche Feldtheorie/Einheitliche Feldtheorie und Hamiltonsches Prinzip", *Sitzungsberichte der Preussischen Akademie der Wissenschaften zu Berlin*, Physische – Mathematische Klasse, BERLIN: 1929, pp. 2–7/156–159.

p. 201 – IDEM, see also note for p. 186 above.

p. 203 – IDEM, see also note for p. 10 above.

p. 203 – Leon M. LEDERMAN/Dick TERESI, *op. cit.*, p. 21.

p. 204 – Paul Charles William DAVIES, *Superforce. The Search for a Grand Unified Theory of Nature*, LONDON: Heinemann, 1984; NEW YORK: Simon & Schuster, 1964, p. 168.

p. 205 – Rudjer Josip BOSKOVIC, *Theoria philosophiae naturalis redacta ad unicam legem virium in natura existentium*, VENEZIA: Remondini, 1763.

p. 205 – IDEM, *De solis ac lunae defectibus*, LONDON: Millar & Dodsleios, 1760.

p. 205 – See Rudjer Josip BOSKOVIC, *Les éclipses*, French trans. Augustine DE BARRUEL, *De solis ac lunae defectibus* cit., PARIS: Valade/Laporte, 1779, p. XXIX.

p. 211 – Albert EINSTEIN, see also note for p. 185 above.

p. 212 – Paul Charles William DAVIES, *Superforce* cit., p. 89.

p. 213 – Steven WEINBERG, *Dreams of a Final Theory. The Search for the Fundamental Laws of Nature*, NEW YORK: Pantheon, 1993, p. 18; LONDON: Hutchinson Radius, 1993, p. 13.

p. 217 – Carl SAGAN, in Stephen William HAWKING, *op. cit.*, "Introduction", p. x.

p. 218 – Stephen William HAWKING, *op. cit.*, pp. 174, 175.

p. 219 – James JEANS, *The Mysterious Universe*, CAMBRIDGE: Cambridge University Press, 1930; NEW YORK: Macmillan, 1930, p. 134.

p. 219–220 – George SMOOT, *Time*, NEW YORK: 28 December 1992.

p. 220 – Leon M. LEDERMAN/Dick TERESI, *op. cit.*, p. 24.

p. 220 – *Ibid*, Interlude C ("How we violated parity in a weekend . . . and discovered God"), pp. 256–273.

p. 220–1 – *Ibid*, p. 254.

p. 221 – Paul Charles William DAVIES, see note for p. 8 above (*God and the New Physics*).

p. 221 – Paul Charles William DAVIES, see note for p. 8 above (*The Mind of God*).

p. 221 – Frank J. TIPLER, *The Physics of Immortality. Modern Cosmology, God, and the Resurrection of the Dead*, NEW YORK: Doubleday, 1994; LONDON: Macmillan, 1995, p. 1.

p. 221 – John C. POLKINGHORNE, *The Faith of a Physicist*, PRINCETON (NJ): Princeton University Press, 1994.

p. 222 – See Robert John RUSSELL/W. R. STOEGER/G. V. COYNE (eds), *Physics, Philosophy, and Theology. A Common Quest for Understanding*, CITTA DEL VATICANO/NOTRE DAME (Ind.): Vatican Observatory/University of Notre Dame Press, 1988; – IDEM/Nancey MURPHY/ C. J. ISHAM (eds), *Quantum Cosmology and the Laws of Nature. Scientific Perspectives on Divine Action*, CITTA DEL VATICANO: Vatican Observatory and the Center for Theology and the Natural Sciences, 1991.

Chapter 10. The Ascent of Mathematical Woman

p. 223 – Fay AJZENBERG-SELOVE, *A Matter of Choices. Memoirs of a Female Physicist*, NEW BRUNSWICK (NJ): Rutgers University Press, 1994, p. 91.

p. 224 – *Ibid*, loc. cit.

p. 225 – Robert OPPENHEIMER, quoted in Sharon BERTSCH MCGRAYNE, *op. cit.*, p. 264.

p. 226 – Sharon BERTSCH MCGRAYNE, *op. cit.*, p. 277.

p. 228 – Harriet ZUCKERMAN/Jonathan R. COLE/John T. BRUER (eds), *The Outer Circle. Women in the Scientific Community*, NEW HAVEN (Conn.): Yale University Press, 1992, p. 13.

p. 229 – Myra SADKER/David SADKER, *Failing and Fairness. How America's Schools Cheat Girls*, NEW YORK: Scribner, 1994.

p. 230 – *Ibid*, pp. 1–2.

p. 230 – "This is what you thought: were any of your teachers biased against females?", *Glamour*, NEW YORK: August 1992, p. 157.

p. 230 – See Mary Frank FOX, "Gender, environmental milieu and productivity in science", in Harriet ZUCKERMAN/Jonathan R. COLE/John T. BRUER (eds), *The Outer Circle* cit., pp. 191–192, 195–196.

p. 230 – Sharon TRAWEEK, *Beamtimes and Lifetimes. The World of High Energy Physicists*, CAMBRIDGE (Mass.): Harvard University Press, 1988, p. 116.

p. 230 – Fay AJZENBERG-SELOVE, *op. cit.*, p. 209.

p. 230 – Mary Frank FOX, *op. cit.*, in Harriet ZUCKERMAN/Jonathan R. COLE/John T. BRUER (eds), *The Outer Circle* cit., p. 196.

p. 231 – Quoted in Mary Frank FOX, *op. cit.*, in Harriet ZUCKERMAN/Jonathan R. COLE/John T. BRUER (eds), *The Outer Circle* cit., pp. 191–192.

p. 231 – Doreen KIMURA, "Sex differences in the brain", *Scientific American*, NEW YORK: vol. 267, no. 9/September 1992, pp. 118–125.

p. 232 – NATIONAL SCIENCE FOUNDATION, *Women, Minorities, and Persons with Disabilities in Science and Engineering 1994*, WASHINGTON (DC): National Science Foundation, 1994.

p. 233 – Myra SADKER/David SADKER, *op. cit.*, pp. 93–98.

p. 233 – Anne FAUSTO-STERLING, *Myths of Gender. Biological Theories About Women and Men*, NEW YORK: Basic Books, 1985, 1992².

p. 234 – *Ibid*, pp. 54–59.

p. 234 – J. Mck. CATTELL/D. R. BRIMHALL (eds), *American Men of Science. A Biographical Dictionary*, LANCASTER, (PA): Science Press, 1921.

p. 236 – Sharon TRAWEEK, *op. cit.*, pp. 29, 83.

p. 238 – Leon M. LEDERMAN, "Science and the bottom line", *The New York Times*, NEW YORK: 16 September 1993.

p. 239 – Robert Rathbun WILSON, conversation with Senator John PASTORE, quoted in Leon M. LEDERMAN/Dick TERESI, *op. cit.*, p. 199.

p. 242 – See Evelyn FOX KELLER, *Reflections on Gender and Science*, NEW HAVEN (Conn.): Yale University Press, 1985.

p. 242 – Karen BARAD, "A feminist approach to teaching quantum physics", in Sue Vilhauer ROSSER, *Teaching the Majority. Breaking the Gender Barrier in Science, Mathematics, and Engineering*, NEW YORK: Teachers College Press, 1995, pp. 43–75.

p. 243 – Margaret WERTHEIM, "Falling for the Stars", interview with Sandra FABER, *Vogue Australia*, GREENWICH: March 1993, p. 90.

p. 243 – Corey S. POWELL, "Profile: Jeremiah and Alicia Ostriker" *Scientific American*, NEW YORK: vol. 271, no. 9/September 1994, pp. 28–31.

p. 243 – Lynn MARGULIS, *Symbiosis in Cell Evolution. Life and Its Environment on the Early Earth*, NEW YORK/OXFORD: Freeman, 1981; also *Symbiosis in Cell Evolution. Microbial Evolution in the Archean and Proterozoic Eons*, NEW YORK: Freeman, 1992, 1993².

p. 243 – Dian FOSSEY, *Gorillas in the Mist. A Remarkable Woman's Thirteen Year Adventure in Remote African Rain Forests with the Greatest of the Great Apes*, BOSTON: Houghton Mifflin, 1983; LONDON: Hodder and Stoughton, 1983.

p. 244 – Adrienne ZIHLMAN, "Women and evolution", *Signs. Journal of Women in Culture and Society*, CHICAGO: IV/1978, pp. 4–20;

– "Pygmie chimpanzee morphology and the interpretation of early hominids", *South African Journal of Science*, PRETORIA: vol. 75.4/1979, pp. 165–168.

p. 244 – Evelyn FOX KELLER, *A Feeling for the Organism. The Life and Work of Barbara McClintock*, SAN FRANCISCO: Freeman, 1983, ch. XII ("A feeling for the organism"), pp. 197–207.

p. 244 – James E. LOVELOCK, *Gaia. A New Look at Life on Earth*, OXFORD: Oxford University Press, 1979.

– *The Ages of Gaia. A Biography of Our Living Earth*, OXFORD: Oxford University Press, 1988.

p. 246 – Londa SCHIEBINGER, interviewed by Margaret WERTHEIM.

p. 247 – Evelyn FOX KELLER, "The wo/man scientist", in Harriet ZUCKERMAN/Jonathan R. COLE/John T. BRUER (eds), *The Outer Circle* cit., p. 235.

p. 247 – Quoted in Sharon TRAWEEK, "High energy physics: a male preserve", *Technology Review*, CAMBRIDGE (Mass.): vol. 42/ November–December 1984.

p. 247 – Evelyn FOX KELLER, "The wo/man scientist" cit., in Harriet ZUCKERMAN/Jonathan R. COLE/John T. BRUER (eds), *The Outer Circle* cit., pp. 234–235.

p. 248 – Stephen William HAWKING, *op. cit.*, pp. 174–175.

p. 250 – Albert EINSTEIN, see. note for p. 12 above.

p. 250 – Michael J. BUCKLEY, "The newtonian settlement and the origins of atheism", in Robert John RUSSELL/William R. STOEGER/George V. COYNE (eds), *Physics, Philosophy, and Theology. A Common Quest for Understanding* cit., p. 99.

p. 250 – Steven WEINBERG, *op. cit.*, pp. 244–245.

p. 250 – Michael J. BUCKLEY, *op. cit.*, p. 99.

p. 251 – Evelyn FOX KELLER, *Secrets of Life, Secrets of Death. Essays on Language, Gender and Science*, NEW YORK/LONDON: Routledge, 1992, p. 5.

INDEX